THREE YEARS
in the
Army of the Potomac

by Henry N. Blake

Eleventh Regiment
Massachusetts Volunteers

Edited by Robert D. Morss
Foreword by Frederick B. Arner

Cross Sword Press
Sanford, Maine
2026

Published by Cross Sword Press
Sanford, Maine.

Footnotes and Appendix copyright © 2026 Robert D. Morss.
All rights reserved.

Blake, Henry Nichols, 1838-1933.
Three Years in the Army of the Potomac (Annotated) / by Henry N. Blake :
edited by Robert D. Morss : foreword by Frederick B. Arner.

Originally published : Boston : Lee and Shepard, 1865.
Includes bibliographical references.
ISBN 979-8-9955454-0-8

Cover: "The Battles Before Richmond – Battle of Malvern Hills, near Turkey
Bend, James River, Va., fought Tuesday, July 1 – Final repulse of the rebels, 5
o'clock P.M. – From a sketch by our special artist, Mr. William Waud" *Frank
Leslie's Illustrated Newspaper.* August 2, 1862.

Table of Contents

Dedicated to the memory of

Captain Henry N. Blake
General William E. Blaisdell
11th Massachusetts Volunteer Infantry
and
Frederick B. Arner

Foreword

Three Years in the Army of the Potomac was written in late 1864 and early 1865 while its author, Henry Nichols Blake, was recuperating from a wound he received at the Battle of Spotsylvania Court House. The book constitutes a history of the 11th Massachusetts Volunteer Infantry for three of its four-year existence, but limited to the perspective of a line officer who had a sharp sense of situational awareness, and kept a detailed diary. It differs from most unit histories in its temporal immediacy, published just as the war was ending, with the specific intent of exposing certain aspects of the military establishment and, in particular, the shortcomings of a number of Union generals Blake felt he was "compelled to serve under." It is unique in ways which reflect the author's personal characteristics and political beliefs. Blake candidly offers his opinions as unvarnished truth, but manages to do so without appearing smug or self-congratulatory. Moreover, his omission of significant personal details, such as his wounds and promotions, and his oblique references to himself and other members of the regiment, give the book a remarkably impersonal quality. For instance, he does not mention that his younger brother William had enlisted along with him into Company K and died of disease in December 1862, while the regiment was encamped near Fredericksburg. Nor does he relate that his commanding officer, Col. William Blaisdell, covered up for Blake's absence-without-leave to accompany William's body back to Boston for burial.

However, when it comes to outlining perceived high or low level corruption or mismanagement, Blake gets very personal and strident. Captain Blake was writing as an early "whistle blower" to get a few things off his chest. This is recognized by the only contemporary notice we have found – the *Army and Navy Journal* of May 13, 1865 (Vol. II, No. 38) commented on the unusual nature of the work:

> *It is the boldest and bitterest stricture on military operations as yet evoked by this war — a spoonful of fresh horseradish, quite a relish in its peculiarity. If sustained by evidence, it will prove an important source of information for writers upon the operations of the Army of the Potomac. Without intending to express any decision upon its historical merits at the present time, or upon its errors — it is clearly and agreeably written, and well worthy of examination, if only as a curiosity in the military literature of the war. The author, we are told, is a brave and capable man, keen, observant, chiefly celebrated in the army as a chess player, and judging from his career in arms, irrespective of that with his pen, one of the most fearless who has ever undertaken to expose what he deemed abuses and defects in military movements, administrations and men.*

The book is enigmatic for the modern reader. While Blake states in his preface the names of the general officers that are the subjects of his greatest criticism, they are not always specifically named when censured in the text. Blake's target audience of 1865, the officers and men of the Army of the Potomac, would have easily recognized the subject of each insult, but with the passage of a century and a half, a little detective work is required — and Blake usually leaves enough clues to figure out each allusion. Although he denied that this approach was based upon fear of litigation, he does appear to be giving himself just enough distance to avoid it.

In the book, however, Blake makes no mention of his chess playing, or how he otherwise spent his spare time. Literary allusions are sprinkled throughout the text, but this is no surprise, given his educational background. A photograph from the period shows the newly-commissioned Henry Blake at age twenty-three in his lieutenant's uniform, bespectacled, and chin-whiskered – not a particularly dashing officer who, in modern parlance might be considered somewhat of a nerd. However, he joined the first company formed in his home town of Dorchester immediately after the fall of Fort Sumter.

The 11th Massachusetts (or as Blake later referred to it, the "lousy 'leventh") was a plebeian regiment, and one might have expected that Blake, as an alumnus of Harvard, would have advocated for the Dorchester company to apply for membership in the 20th Massachusetts, the so-called "Harvard Regiment," for the potentially greater intellectual stimulation of its officer corps. He would have known, however, as a staunch Republican, that he would encounter considerable opposition to his abolitionist views in the 20th Massachusetts, which was jokingly called a "Copperhead Regiment" for the number of Democrats among its officers.

Our knowledge of the personal interplay between Blake and the other more blue-collar officers and men of the 11th Massachusetts comes almost entirely from anecdotes in his post-war writings. Although he appears to have been instrumental in getting the Dorchester Company accepted into the 11th Massachusetts, Blake declared in his memoirs that in the election of officers, he did not obtain a single vote: "Men who had run with the [beer] tub were in the majority." Appointed Sergeant in Company K, and promoted to Second Lieutenant within a year, Blake was not the typical junior officer of a regiment whose officers had a dismissal and resignation rate of major proportions. His military records include a request for leave in Washington so that he could "visit the halls of Congress and the Supreme Court," and every indication is that he did precisely that. With most other officers of the 11th, such a request would have been met with eyebrow-raising skepticism if not outright laughter.

While there is little information in Blake's book about Col. William E. Blaisdell, the 11th Massachusetts' commanding officer for most of the war, more is contained in the annotations (and the Appendix), drawn from Blake's unpublished memoirs. History's view of Blaisdell is dominated almost exclusively by the published letters of his long-term adversary in the army, the puritanical Col. Robert McAllister, of the 11th New Jersey, who did not see many redeeming characteristics in the Massachusetts colonel who was well-known in the army for his liberal use of alcohol and profanity.

Blake starts his narrative on July 16, 1861 as the Union army sets out on the campaign that results in the first battle of Bull Run. For the initial three months of the regiment's history — April to July 1861, during which it was recruited and trained in Massachusetts and had a somewhat adventurous train journey to Washington — we have included an excerpt from Blake's unpublished 1916 memoirs. A much shorter history of the 11th Massachusetts by Sgt. Gustavus B. Hutchinson, Company D, was published in 1893 for the benefit of the veterans association. This 96-page book provides excellent details on the formation of the 11th in 1861, but adds little to the Bull Run to Spotsylvania period that Blake covers in *Three Years in the Army of the Potomac*. Hutchinson, who was discharged for disability in May 1863, relies almost entirely on Blake, and Martin L. Haynes' history of the 2nd New Hampshire, which was in the same brigade as the 11th Massachusetts from October 1861 through the Battle of Chancellorsville in May 1863. Hutchinson also provides a very sketchy account of the 11th Massachusetts for 1864 and 1865, freely adapted from the reports of the Massachusetts Adjutant General. For this latter period, the letters of General Robert McAllister, who was the 11th Massachusetts' brigade commander for most of late 1864 and 1865, becomes an essential source, as does Sergeant Thomas D. Marbaker's history of the 11th New Jersey, which was in the same brigade as the 11th Massachusetts from Fredericksburg through Appomattox, except for the period April through July 1864, when the 11th Massachusetts was part of the "Excelsior Brigade" of New York regiments.

After the three-year veterans went home in June 1864, the 11th Massachusetts was a wholly different unit. Three-hundred sixty-three reenlisted veterans, transfers, and recruits remained — enough to retain the unit identity as the 11th Massachusetts Battalion. Whatever remained of unit cohesion was diluted by the large influx of soldiers from other regiments. Some forty-five enlisted men of the expired 1st Massachusetts had joined the 11th after Spotsylvania, and when the 16th Massachusetts' term expired in July 1864 its 190 remaining personnel were incorporated as two companies of the 11th. For the final ten months of the war (Petersburg to Appomattox), the 11th Massachusetts was an over-sized battalion of seven companies. The recuperating Captain Blake contemplated rejoining the battalion, but after his friend, Captain Alexander McTavish, was killed at Boydton Plank Road in late October 1864, he thought better of it. Both Blake and McTavish had been wounded at Spotsylvania and returned to Boston together intending to return to the front upon recovery even though their enlistments were up.

Three Years in the Army of the Potomac contained no pictorial material; we have included photographs of Blake and

Blaisdell, and a brief biography of each. Footnotes are presented to verify, amplify, and clarify various aspects of the text. They contain material Blake felt was inappropriate for publication in 1865, but did include in his unpublished memoirs, kindly provided by his descendants. Written in 1916 and based on scrapbooks he kept during his long life, Blake's memoirs fill in many of the blanks in the history of the 11th Massachusetts. They also provide biographical information that has been used in articles about him published by Harvard Law School, and in Montana, his adopted residence for four decades following the war.

Frederick B. Arner
Kensington, Maryland (2002)

Editor's Note

This annotated reprint of *Three Years in the Army of the Potomac* started as a joint project between Frederick B. Arner and myself in 1997, based on a common interest in the Eleventh Massachusetts – each of us having an ancestor in its ranks.

Mr. Arner's grandfather was Pvt. Samuel L. Blood, who was recruited at age 18 as a replacement drummer just before the 1864 campaign — no doubt enticed by the $325 bounty. Pvt. Blood was wounded at Spotsylvania and discharged after only a few weeks in the regiment. For me, it was my grandfather's uncle, Pvt. Walter S. Morss, who enlisted, also 18 years of age, in the original 1861 muster of the 11th and served his three-year term as clerk of Company G without serious injury. (He spent a few weeks in mid-1862 as a Confederate prisoner.)

Mr. Arner, having published two books on Army of the Potomac politics, had started writing a new regimental history of the 11th Massachusetts, and shared with me an early draft of his introduction and first two chapters. I noted that since Blake, of necessity, was massively quoted, why not just issue a reprint of Blake, with footnotes to assist modern readers through the often obscure allusions. This plan was agreed, and over the next several years we developed a productive research collaboration by postal mail, and eventually email (these were early internet days). Our goal with this edition has been to provide a modern reprinting of the original text, fleshed-out with annotations giving further insight into the character of its author and his time in the army. The result is something of a companion volume to *The Mutiny at Brandy Station* (1993), the story of the Third Corps break-up, in which Henry Blake was entangled. Indeed,

Mr. Arner's research for that book formed the basis for our annotated Blake project.

Since Blake's narrative has specific starting and ending dates — July 16, 1861 to May 12, 1864 — we considered the need for supplementary material about the time periods in the regimental story not covered by Blake. Having enlisted in the first Dorchester company formed within two days after the fall of Fort Sumter, Blake was present from the very beginning of the 11th Massachusetts. Because of this, it seemed appropriate to include his 1916 memoir, "Rumors of War," touching on this early period. Conversely, we have omitted coverage of the post-Blake period: June 1864 to May 1865. We considered including an extended epilogue, but with Blake convalescing at home, Blaisdell killed in action, and the massive turnover of personnel, the regiment had become the 11th Massachusetts in name only. It remained with the Second Corps through the end of the 1864 Overland Campaign (spared the Cold Harbor debacle), the ten-month siege of Petersburg, and the final pursuit to Appomattox Court House. Before being mustered out the 11th participated in the Grand Review of the Armies in Washington, May 1865.

The intent of the footnotes is to provide context, identify historical, literary, and cultural allusions, and unpack Blake's sometimes ungainly sentence structure. We have not attempted to fully explain each battle – Blake himself noted that his narrative was limited to what he himself saw and heard. However, the reader will find an increase in footnoting for the battles in which the 11th Massachusetts was heavily involved in combat – it was in every campaign of the Army of the Potomac except Antietam.

Articles in the Appendix expand on significant events for the 11th that Blake glosses over. Blake's immediate concerns in 1864-5 seem to have been his dealings with the army medical department and his apparent eagerness to finally release his pent-up ire in print against the generals he judged to be grossly incompetent – the book was published within a year following Spotsylvania. Despite stating in his memoirs that he had intended to return to the battalion once his wound healed, the book conveys an ever-increasing sense of exasperation. Disgusted by his April 1864 court-martial over the Third Corps breakup, and his three years of putting up with what he considered to be high-level military incompetence, it seems likely that he'd had enough of the army.

The text is presented as it appears in the original 1865 edition, retaining spellings then in common use – which today would be corrected by a spell-checker – such as "Spottsylvania," "Kearney," and "reconnoissance." Blake spelled "negro" uncapitalized in 1865, but capitalized it in his memoirs. "Rebel," on its own, was lower-case; but capitalized when speaking of "the Rebellion." Obvious typesetting errors have been corrected and footnoted. Chapter titles have been amended slightly for brevity. The only photographs we have included, with biographical notes, are those of Blake and Blaisdell. The latter because, unlike the ever-changing brigade, division, corps, and army commanders, Blaisdell's leadership position in the 11th Massachusetts lasted for the entirely of Blake's three years in the Army of the Potomac.

Robert D. Morss
Sanford, Maine (2026)

Unit Designations

In the footnotes and appendix a Union formation's place in the command structure of the army may appear in an abbreviated format: Arabic numbers for a brigade or a division; Roman numerals for a corps. For example, the 11th Massachusetts, for most of its service, was in the 1st Brigade of the 2nd Division of the Third Corps, which is abbreviated as 1/2/III. At the Wilderness and Spotsylvania it was in 1/4/II. These designations sometimes appear alongside the name of a unit's commanding officer.

The following is a list of the brigades, divisions, and corps to which the 11th Massachusetts was assigned during Henry Blake's service with the regiment.

1861: Bull Run

- 1st Brigade: Franklin
- 3rd Division: Heintzelman
- Army: McDowell

1862: The Peninsula and 2nd Bull Run

- 1st Brigade: Naglee (Yorktown), Grover (Williamsburg *et seq*)
- 2nd Division: Hooker
- III Corps: Heintzelman
- Army: McClellan, Pope

1862: Fredericksburg

- 1st Brigade: Carr
- 2nd Division: Sickles
- III Corps: Stoneman
- Army: Burnside

1863: Chancellorsville

- 1st Brigade: Carr
- 2nd Division: Berry
- III Corps: Sickles
- Army: Hooker

1863: Gettysburg

- 1st Brigade: Carr
- 2nd Division: Humphreys
- III Corps: Sickles
- Army: Meade

1863: Mine Run

- 1st Brigade: Blaisdell
- 2nd Division: Prince
- III Corps: French
- Army: Meade

1864: The Wilderness and Spotsylvania

- 1st Brigade: McAllister
- 4th Division: Mott
- II Corps: Hancock
- Army: Meade/Grant

Rumors of War

April – July 1861

ABRAHAM LINCOLN was elected President November 6, 1860, and rumors from the South filled the air, when the result was proclaimed, and the advocates of secession (another name for treason) took bold steps destined to end in blood. Many sincere citizens did not believe it possible that armies could be raised in the abodes of slavery to fight under the standards of disunion and fire on the American flag.

The bombardment of Fort Sumter in April, 1861 changed the situation in an instant, sight was restored to the blind, and the loyal people awakened from lethargy. Even then, there were intelligent men who protested there would not be any war, that time and money were wasted in equipping regiments and others who were positive the troubles would end in sixty or ninety days. I knew Harvard students from the seditious states who inherited the malignant doctrine of state rights and were zealous to vindicate secession by an appeal to arms.

Captain Benjamin Stone, Jr., of the Massachusetts militia, who taught a singing school I attended in the Harrison Square Church[1] prepared, April 18, 1861, a call for volunteers, and I signed the paper within forty-eight hours in Lyceum Hall, Dorchester. I promised to enlist in the company for the impending war, which I was convinced was to involve our country, although its duration was beyond my dreams.[2] My

---

1 The name "Harrison Square" commemorates a visit to Dorchester by William Henry Harrison during the presidential campaign of 1840.

2 Blake later noted: "I was not alone, however, and my brother, William Edward Blake, signed the enlistment papers of Captain Stone – a mere youth of military age."

business was abandoned, my mind was excited by the crisis, and my time was devoted to drills in the manual of arms in Lyceum Hall and Fort Warren. I do not believe there was any person who was more ignorant in the details than myself. I despised the militia that appeared to be organized for fun and conviviality, and did not know how to handle or load a gun. The first call of President Lincoln was for 75,000 volunteers, and the young men responding exceeded this number and it seemed possible the Dorchester company might be disbanded because its services were not needed.

My neighbor, George Clark, Jr., of Harrison Square, a splendid drill master in the militia, was raising a regiment, and had eight Boston companies, and I interviewed him without any authority. He informed me he had received applications from several parties to furnish companies for the two vacancies, and had not acted on them, and finally agreed to receive our company with his command. I made a report on this proposition at the next drill meeting and on my motion, the members voted favorably, and thus became in history, Company K, 11th Regiment, Massachusetts Volunteers Infantry. An informal election of officers was held in Lyceum Hall at which a captain and four lieutenants were selected and I did not obtain a vote.[3] Men who had run with the tub were in the majority, and I was not a candidate for anything, and one of the gang told me two years afterwards he believed I was after a commission and hence was surprised when I shouldered my musket in the ranks. This candid and gallant soldier who fell in the color guard at Gettysburg was

---

3   The company officers were elected in the last week of April and on May 1st they met at the Parker House and chose the following regimental field officers: George Clark Jr. (Colonel); William Blaisdell (Lieutenant-Colonel); George F. Tileston (Major). Hutchinson says Clark and Tileston were connected to the Boston press and were able to secure financial support required by the regiment until it was mustered into Federal service.

not alone in his opinion, and his statement was uttered in a tone which indicated a confession of injustice to me and my motives. I answered truly that the last thing I thought of at the beginning of hostilities was military rank, and I was proud to serve as a Private. I was mustered into the service as Second Sergeant of the Company, without any request on my part, or any person in my behalf. I was the choice of Captain Stone.

The company was sent May 27, 1861, to Fort Warren[4] and formed the tenth in the order of alignment in the regiment, and the comforts of home were exchanged gladly in a spirit of patriotism for the easements of the fortress. I was designated by Captain Stone to act as a Sergeant, and it was my duty to go to the headquarters of the regiment at a certain time each morning and copy from the order book of the adjutant all orders issued by the commander of the regiment. We lived in a new world and hours for drill and fatigue work were prescribed and enforced.

A momentous and historic event was recorded June 13, 1861, in Fort Warren when the regiment was mustered into the service of the United States for the term of three years unless sooner discharged by the end of the war. A physical examination took place prior to the departure for Fort Warren by a physician of our own family, and I knew my eyes had not been tested according to the strict requirements of the regular army, and I adopted the safe course of keeping my spectacles in my pocket while the official oath was administered. I may digress at this stage and say I was born near-sighted like my sister, and used glasses five years before I enlisted, and was as efficient in my

---

4   Fort Warren, Georges Island, Boston Harbor, was used as the assembly point and initial training ground for the 11th, 12th, and 14th Massachusetts. Soon after the first Bull Run battle it was used as a prison for Confederate soldiers.

military career as if I enjoyed perfect vision. I tried to keep on hand two pair of specs so that I would be prepared for accidents.

The Eleventh and Twelfth Massachusetts Regiments, Infantry, were stationed at Fort Warren and engaged early and late in preparedness. The Eleventh left Fort Warren June 17, 1861, and proceeded to North Cambridge[5] where it encamped at Camp Cameron, so named in honor of a Secretary of War. The rations furnished to the regulars by the Government were issued to the Eleventh on this day for the first time. The routine of drills by squads, drills by companies, drills by battalions and drills by the regiment continued until June 28 when the last dress parade took place. A state flag was given to the Eleventh and my friend, Hon. Richard H. Dana,[6] was the speaker of the occasion. Mrs. E. H. Sanford[7] presented by the bearer, her son, J. E. M. Sanford, thirteen years old, the national colors of her own design.[8]

---

5 "Up the harbor [by steamer], along the Charlestown shore, the celebration of the battle of Bunker Hill was in progress. When the regiment landed it marched through Boston to Winthrop, where a collation was served at the expense of the city of Charlestown." [Hutchinson]

6 Richard Henry Dana was a writer, lawyer, and anti-slavery activist, known to most people at that time for his sea travels, which he wrote about in *Two Years Before the Mast* (1841).

7 Edward H. Sanford was a ship's captain whose father had founded the Boston and Bangor Steamship Company. The company offered Gov. Andrew free passage for troops and supplies as a patriotic gesture. By early 1862 the Federal government had requisitioned all but one of the company's ships. [*History of the Boston and Bangor Steamship Company,* 1882]

8 *Boston Daily Courier* noted that Mrs. Sanford's flag was one and one-half yards square, cost $200, and was "composed of double silk, one side white, the reverse red, on the white side a beautiful painting of the Goddess of Liberty; above it in a gilt scroll '11th Regiment M. V. M.' under it gold letters, the following motto, 'Constitutional Rights.' On the reverse, thirty-four gold stars, in form of a star; above, the motto 'Three cheers for the Union;' below the stars, the date. The banner was trimmed in heavy gold fringe, gold cord and tassels." In 1916 Blake noted that the Sanford banner had been "recently returned" to the son, "a good friend of the survivors of

At 2 o'clock the Eleventh bid farewell to Camp Cameron, June 29, 1861, and marched to the Boston station of the Old Colony Railroad and boarded the cars for Fall River. The streets on the route of the march were lined with cheering sympathizers, "good bye" was spoken to weeping relations and dear friends, and in some instances it was the last farewell. The 11th embarked on a steam boat at Fall River, and after an uneventful passage in the night, arrived in New York City at 10 o'clock A.M. and marched down Broadway.[9] Not one of the 978 musket bearers of the 11th was missing June 30, but this remark does not apply to the towels which were seen in Virginia months later, bearing the name of the boat stamped thereon.

I am writing of what occurred fifty five years ago when railroading was in an infant stage and the long trains were impeded by necessary delays, but the states of New Jersey, Pennsylvania, Delaware and Maryland were crossed in safety.[10] The 11th passed through Baltimore at 6 o'clock, A.M. July 2,

the Eleventh." Its status today (2025) is unknown.

9 "[We were] escorted to the barracks at City Hall Park [and] formally welcomed by Richard Warren, Esq., a fitting response being made by Col. Clark and Surgeon L. V. Bell. The regiment then filed into the barracks for refreshment, while the officers were entertained at the Astor House." [Hutchinson] Richard Warren was president of the Sons of Massachusetts in New York City. The surgeon of the 11th, Luther V. Bell, had a long career in the treatment of mental illness. By the time of his death in early 1862 he was surgeon for Hooker's entire division. Warren's speech, and Bell's reply were printed in full in the *New York Times*, July 1, 1861.

10 Hutchinson continued: "[The] regiment embarked on the steamer *Kill von Kull* for Elizabethport [New Jersey], from which they proceeded [by rail], via Harrisburg, to Baltimore and Washington. Soon after leaving Elizabethport, one of the engines left the track, endangering the whole train and its more than one thousand passengers. Through the assistance rendered by expert railroad men belonging to the regiment, the engine was soon replaced upon the track," The July 1, 1861 *New York Times* noted: "It is rumored that the reason they were sent by this roundabout way was because a member of the Cabinet has stock in the railroad."

1861, ready for an attack with muskets half cocked, but was not molested. After the assault on the Sixth Massachusetts Infantry [April 19], the men were drilled daily in street firing tactics which are effective against a mob without discipline.[11] It is known at this time, however, that most of the assailants in this affair in Baltimore were recruits for the rebel forces, acting under orders of officers and prepared for the dastardly work.

The 11th halted in Washington at noon, July 2, and marched from the railroad station down Pennsylvania Avenue and by the White House where stood Abraham Lincoln, greeting the troops answering the call of the President.[12] Hearty cheers for Lincoln resounded through the ranks of the 11th and the tents were pitched on land between the White House and the Washington Monument, then less than 100 feet in height.[13] An order was issued July 3 by Col. Clark naming the ground Camp Sanford, in honor of Mrs. E. H. Sanford, and there were in the 11th, 1,048 officers and enlisted men. I availed myself of the opportunity to visit objects of interest and rushed to the Capitol to see the

---

11 "Col. Clark, having been apprised that he might anticipate serious trouble in Baltimore, just before reaching that city ordered the muskets to be loaded, and made every preparation to repel [an] attack and to force a way through the city in case of resistance." [Hutchinson]

12 Hutchinson noted a laudatory appraisal of the regiment which appeared in a Washington paper: "Unquestionably one of the finest regiments that has yet marched through the avenue ... headed by Gilmore's celebrated band, and followed by twenty large new covered wagons, each filled with camp equipments, etc., and drawn by four fat and sleek matched horses. The Eleventh in their march up the avenue, were pronounced a little in advance of the best A.1. regiments yet in the line of arrival." Patrick S. Gilmore was an Irish-born bandmaster and composer who had founded "Gilmore's Band" in 1858. The band accompanied the 11th from Boston. It later enlisted *en masse* as the band of the 24th Massachusetts.

13 The Washington Monument eventually rose to 555 feet, but it was not until 1888 that the capstone was placed.

Supreme Court and Congress in session.[14] Ex-Vice President Breckinridge sat in his seat as a Senator from Kentucky and resigned October 8, 1861, and joined the Confederate Army.

I soon discovered the Capitol was "in the enemy's country," and both sexes, especially the female, showed utter contempt for the Union cause by their acts, and discreetly maintained silence in the presence of soldiers. I noticed in the minstrel shows the jokes, pleasing the crowd, were hostile to "niggers" and their white "worshipers" and favored the southern view of the struggle. The bar rooms were thronged by the commissioned officers, but orders from headquarters deprived the privates of the glorious liberty of getting drunk. I approved one side of the military life in Washington – the enforcement of orders of this nature – and a drunken enlisted man was rarely seen.

The 11th crossed the Rubicon[15] (Potomac) July 14, 1861, and camped on Shuter's Hill, near Fort Ellsworth, Alexandria, Virginia. The new camp was named Wilson as a compliment to the Senator from Massachusetts. The 11th was assigned to a brigade with the Fifth Massachusetts, Fourth Pennsylvania, First Minnesota and Rickett's Battery, and went forth in battle array July 17.

Henry N. Blake
Boston (1916)

---

14 A 22-year-old, Harvard-educated attorney, Blake was not the typical enlisted man. In addition to his temperance and his abolitionist leanings, he was keenly interested in politics and government.

15 The point of no return – alluding to Julius Caesar's return to Italy in 49 BCE precipitating the Roman civil war.

THREE YEARS

IN THE

ARMY OF THE POTOMAC.

BY

HENRY N. BLAKE,

LATE CAPTAIN IN THE ELEVENTH REGIMENT MASSACHUSETTS
VOLUNTEERS.

"From year to year, the battles, sieges, fortunes,
That I have passed!"

BOSTON:
LEE AND SHEPARD,
(SUCCESSORS TO PHILLIPS, SAMPSON & CO.)
1865.

Facsimile of the original title page.
Quotation is from Othello, *Act I, Scene iii*

Preface

THE AUTHOR formed a thousandth part of the Eleventh Regiment Massachusetts Volunteers, and enlisted in April 1861, and was mustered out of the service in June, 1864. During this period, he recorded in a diary every incident of interest which passed under his observation; and the request of many comrades, who saw him take notes upon the march or on the battle-field, induced him to prepare them for general reading. In carrying out this design, the author has most unwillingly omitted to mention by name the officers and enlisted men of his regiment, although some of their heroic deeds are briefly described.

Rarely venturing to go beyond his limited vision, adhering most rigidly to fact, able to prove many strange statements by the testimony of thousands of soldiers, he has found it necessary to portray certain generals and other officers in the strongest colors of shameful cowardice, drunkenness, and military misconduct. The vexations, not the results of litigation, have deterred him from furnishing the names of these obnoxious persons; but, to shield honorable men from base suspicion by civilians, he states that the generals who are censured, with the exception of the foreigners at Fair Oaks and Chancellorsville, one at Malvern Hill second, and one at Bull Run second, are mentioned upon some page of the book. The death of important witnesses might cause a slight difficulty in proving, by direct evidence, incidents which were seen by a very small number of spectators. The author has not attempted to give a history of most of the great engagements in which he bore a humble part; and criticizes only those battles which took place in open country, where he could behold the chief movements of the Union forces.

He has perused many works which have been published upon the present war by quartermasters, chaplains, and correspondents of newspapers, — a class of non-combatants that usually narrate what was observed by others. The author considers that the facts which he has described in the succeeding chapters possess the advantages of originality and reliability. Although the author is aware that we are all prone to error, he has no hesitation in staking his reputation for veracity upon the truth of the statements of this work, however improbable they may appear to the reader.

The names of the following generals, under whom the author was compelled to serve, are not always mentioned when their conduct is described, — IRWIN MCDOWELL, WILLIAM B. FRANKLIN, ALFRED A. HUMPHREYS, WILLIAM H. FRENCH, JOSEPH B. CARR.[1]

---

1 Blake misstates the first names of Irvin McDowell and Andrew Humphreys.

Chapter I

The First Battle of Bull Run

ON JULY 16, 1861, the Eleventh Regiment Massachusetts Volunteers formed a part of a brigade commanded by Col. Franklin,[1] and a division commanded by Col. Heintzelman. In compliance with orders, the regiment marched from Alexandria at two, P.M., and left all the diseased and feeble in camp under the charge of a sick captain,[2] to guard the tents and knapsacks of the men during their absence. The soldiers composing the expedition displayed the highest emotion of joy; and those who were compelled by their physical weakness to remain in the rear were affected with grief, and some shed tears. Each person carried his musket and equipments, containing forty rounds of ammunition; and bore upon his shoulder a woolen blanket enclosed in one made of gum or rubber, and a canteen and haversack. The latter was filled with rations for three days, which consisted of three or four pounds of salt pork or beef ("junk"), thirty crackers ("hard tack"), and a small quantity of sugar and coffee. No one seemed to be informed concerning the object of the events of the uncertain future. The column marched over a narrow and miserable road (one of the chief features of the barbarism of Virginia) south of the Orange and Alexandria Railroad, and formed the left wing of the invading army, which was commanded by Gen. McDowell. Sixteen horses could not

---

1　The brigade also had the 1st Minnesota, 5th Massachusetts, and 4th Pennsylvania. The latter had departed July 20 at the expiration of their three-month enlistment; its commander Col. John Hartranft, remained as an aide to Franklin. The 5th Massachusetts' term had also expired, but the men voted to remain until the conclusion of the campaign.

2　Ten men from each company were detailed as camp guard, under Capt. Benjamin Stone of Company K.

draw a thirty-two pound Parrott gun[3] over the rugged course; and two companies were detached from the regiment to assist the jaded animals in performing this labor. The men sustained the fatigues of their first march during the afternoon and evening in an excellent manner; and there were few cases of utter exhaustion or straggling, although the halts were infrequent. The houses, or, to speak truly, hovels, upon the road, were small in number and dimension, and the country was thickly wooded. The population that was visible comprised aged men, women with their children, and the negroes.

Our progress was extremely slow after sunset; and the column for seven hours advanced, at irregular intervals of time, five, twenty, or one hundred feet. No orders to halt were received during the night from the brigade commander: the delays of a few seconds or minutes were uncertain in their duration; and the men did not know when they could enjoy them. As soon as they had broken ranks, and prepared to rest after a sudden stop, they would be commanded to "fall in" and another pause frequently occurred before the moving mass had traveled the length of a company. The troops, expecting to start at once, sometimes stood in their places half an hour before the march was resumed; and were fatigued during this time, as if they had been in motion. The soldiers were completely exhausted by this severe mode of maneuvering them, for which there was no excuse; and many fell asleep upon the roadside. The regiment reached its halting place near Pohick Church[4] at 3.45, A.M., on [July] 17th, and welcomed

---

3 Part of Battery G, 1st U.S. and called "Long Tom" by its gunners. Lt. Peter Hains, recent West Point graduate, was in charge of it and said it was "drawn by ten horses as green as could be, horses from the farm that had not been trained even to pull together." [Hains]

4 Built in the 1770's, Pohick Church was a landmark due to its association with George Washington. By 1861 it was quite dilapidated, which got worse due to the vandalism of Federal soldiers.

repose without seeking any shelter. A single tree formed the bridge over Pohick Creek, a run which was about twenty-five feet in width, and too deep to be forded; and the troops, assisted by a feeble light, crossed upon it in one rank. The column had been delayed several hours by this obstacle, which could have been easily removed by the pioneers, who carried fifty axes, with which they might have felled the trees that were standing upon the banks of the streamlet, and built a bridge. The most tedious portion of the march could have been prevented by the use of the most ordinary judgment by the brigade commander [Franklin], who displayed a profound ignorance of the first lesson in the school of a general, — the art of marching men: instead of conducting troops a great distance with a small expenditure of strength, he reversed the rule, and caused more fatigue in marching the brigade fourteen miles than they would have suffered in moving twice this distance under an intelligent officer.

The troops rested only an hour, and were awakened at four o'clock,[5] and ordered to resume the march; during which they nibbled their rations, for there were no chances to eat a regular meal. From this early movement until three, P.M., the brigade was marched in the heedless style that characterized the previous night; and no stated halts took place, although there was an intense heat. Hundreds were obliged to leave the ranks, because they had been deprived of bodily vigor by the hardships of the two days; while the brigadier and his staff, riding upon their horses and suffering no inconvenience, unjustly reprimanded them for straggling. These unfortunate soldiers did not wish to avoid the dangers of battle: on the contrary, the only

---

5   If the regiment paused to rest at 3:45 and was awakened at 4:00, this obviously is not an hour. But another member of the 11th recalled a one-hour rest in the *Boston Herald* of July 22, 1861, so 3:45 may be a typo.

apprehension which they expressed was, that the rebels, following the precedent established at Harper's Ferry and Alexandria, might evacuate Manassas. No public road was followed during a portion of the route,[6] which passed through fields and forests in a thinly settled country. The forms of one-half of the brigade arrived at Sangster's Station at three o'clock, about two hours after [Bonham's] South Carolina troops [had] retreated upon the railroad from Fairfax Court House; and the bridges which they had set on fire were burning when the column halted. Squads of the missing fragment of the command constantly joined it during the next six hours, until there were no absentees. Three or four men were killed on both days by the carelessness of soldiers who bore loaded muskets upon their shoulders for the first time.

A drove of pigs, and a flock of sheep numbering about one hundred and fifty, were captured by these men within an hour after their arrival; and it was ascertained that they had been collected for the purpose of feeding a detachment of the rebel army which had been stationed at this point. Some were killed, and roasted upon the camp-fires by means of a rammer, or forked twig, while the flesh quivered. The brigade commander issued an order authorizing the officers to shoot every man who was detected in the act of killing these hogs or sheep, and the soldiers stealthily cooked in the night what they had slaughtered and concealed; but the largest portion of the number was abandoned to nourish the poorly supplied enemy. A circular was transmitted by Gen. McDowell, reproaching the volunteers as plunderers, and denouncing their conduct[7] in such strong terms of

---

6　The march from Alexandria to Pohick Church followed the Telegraph Road, then continued to Sangster's Station over back roads and trails.

7　McDowell stated "with deepest mortification" that he had to reiterate his orders on the preservation of private property. Each regiment was to have

undeserved censure, that a feeling of indignation pervaded the ranks. My facilities for seeing any depredations that might have been committed were excellent, because the regiment had a position in the rear of the division; and the behavior of the troops towards the people upon the road was unexceptionable. A house which had been deserted by the owner, who had joined the forces of Beauregard,[8] was burned during the night by some men who were exasperated on account of the wearisome manner in which they were delayed. They rushed to the wells near some of the dwellings to procure fresh water, because the officer in charge of the command did not halt and allow them to fill their canteens, the contents of which had become too warm for use. Certain mounted officers were very conspicuous in using oaths, and driving the troops from these places which belonged to traitors who were toiling upon the intrenchments of Manassas. The painful experiences and stringent orders of the 16th and 17th excited in the minds of many privates a strong prejudice against some of their superiors in rank, and opinions were freely expressed regarding their wisdom and loyalty. The soldiers listened for the first time to the reports of rebel cannon upon the afternoon of the 18th;[9] and gladly advanced in the direction of the firing at five, P.M., and bivouacked near Centreville at midnight.

Stacks of arms, and batteries, surrounded us in the field near the old road over which Gen. Braddock had led his ill-fated expedition to Fort du Quesne.[10] The army rested two days at this point, and listened to the whistle of the locomotives that were

---

    an officer in charge of a ten-man police force. [OR II, 744]

8   The Confederate Army of the Potomac.

9   The firing they heard was from the skirmish at Blackburn's Ford.

10 The British attempt to capture Fort Duquesne, in western Pennsylvania, during the French and Indian War cost Braddock his life.

bringing to the junction re-enforcements[11] for the rebel hordes. The soldiers eagerly walked long distances to see prisoners; and a defiant [rebel] sergeant told the crowd of spectators that they would "double-quick back to Washington" within a week; a prophecy which was fulfilled upon the 21st. Citizens searching for runaway negroes, or presenting claims for damages to their property, were protected by Gen. McDowell, who allowed them to examine every encampment, and ascertain the number and position of the troops and batteries; after which they rode to Manassas, and communicated the valuable information which they had acquired. A private of the regiment, who was wounded and captured in the battle, saw a person that applied for compensation on account of the injury to his crops, dressed in the nondescript uniform of the Southern soldier. He spoke to him when he was posted upon guard, and asked "How much money did you get for your wheat?" The rebel laughed at the question, but admitted that he was the spy, and entertained his companion by narrating the facts. Rations of pork and beef for two days were boiled on the 20th, and issued to the command at midnight. The regiment was formed in line at one, A.M., upon the 21st: the division commenced to move into the road at half-past two, and marched a mile towards the commanding heights of Centreville, when it was halted to allow the commands of Tyler and Hunter[12] to file by it. Infantry and artillery, during the following three hours, occupied the solitary avenue over which the entire army passed to the front. The appearance of this large force inspired all

---

11 The Army of the Shenandoah under Johnston traveled by train from Front Royal to Manassas Junction to support Beauregard.

12 Only three of the Union army's five divisions were in the Bull Run battle. Tyler's Division feinted the Confederates opposite the Stone Bridge while those of Hunter and Heintzelman crossed Bull Run higher up to flank the Confederate left. Tyler was late getting out of camp, delaying Hunter and Heintzelman.

with confidence; and the order to advance was awaited with impatience.

The head of the column[13] started at the end of this unforeseen delay, advanced upon Warrenton Turnpike through the little village of Centreville, and crossed the bridge that spans Cub Run, near which I noticed about twenty barouches and carriages that contained members of Congress and their friends, who had left Washington for the purpose of witnessing the approaching conflict. The divisions of Hunter and Heintzelman debouched from the main road, at a point two and a half miles from Centreville, and, accompanied by a guide, followed a narrow pathway which was not often used, and led in its tortuous course through a dry territory that was well shaded by the forest. An open space of fifteen acres sometimes intervened; but it was always enclosed by dense woods. The day was one of the hottest of the year: there was no friendly cloud to obstruct the rays of the sun; and it was impossible for the army to march a long distance with unusual speed. Nevertheless, for twelve miles,[14] the men were pushed forward at an unnatural gait, generally walking as rapidly as possible, and double-quicking one-fourth of the time, to keep the different regiments of the column within supporting distance of each other. Nearly every man impatiently asked "How far is it to the Junction?" whenever the loyal citizen residing in the vicinity, who acted as guide, rode along the line. He always answered the question in a good-natured manner by saying, "Six miles." The brigade commander [Franklin] never

---

13 Franklin's order of march was: 1st Minnesota, Ricketts Battery, 5th Massachusetts, 11th Massachusetts. [OR II, 405]

14 McDowell's report says that the longest march was nine and a half miles. Heintzelman was to have forded Bull Run half way to Sudley Ford, but Confederate patrols there would have discovered the flank march, so Heintzelman continued behind Hunter. The head of the column reached Sudley Ford about 9:30 A.M. [OR II, 331]

attempted to secure a rest for the soldiers; and some of them sank upon the ground, wholly overcome by faintness, which was produced by the intolerable heat and the furious rate at which they were marched. There was a very small number, if any, in the Union host, that wished to evade the unknown perils of combat; and many, throwing away their blankets and rations to facilitate their progress, merely retained their muskets and ammunition. The thirty-two pound Parrott gun opened its mouth of iron near the "stone bridge" over Bull Run at 6¼, A.M.; and the artillery upon the left continued to fire at regular intervals in the vicinity of the fords, while the right wing was hastening to turn the left of the rebel line, which was posted in the rear of Bull Run. The scarcity of water to allay the thirst produced by the causes that have been described was another impediment; but the cannonading inspired the men with patriotism, and gave them a physical strength which they could not have possessed under similar circumstances in the avocations of a peaceful life.

They occasionally emerged from the woods, and beheld the long clouds of dust[15] in the south, which showed that the rebels were moving in the same direction; and it required no deep knowledge of mathematics to demonstrate that the two lines of march, if extended, would soon intersect. The column arrived at eleven o'clock at a point that was a short distance from Sedley's Ford; and a slight rest was enjoyed by the brigade while Hunter's division was crossing the run. The smoke of the exploding shells could be distinctly seen. The firing of the infantry and artillery became very active in front, as soon as the advance encountered the rebels, and drove them from their positions. While the men were filling their canteens, an aide-de-camp brought an order

---

15 The dust clouds were caused by the Confederate brigades of Brig. Gen. Barnard E. Bee, Col. F. S. Bartow, and Col. Nathan G. Evans approaching from Manassas Junction.

from Gen. McDowell to send forward two regiments to prevent the enemy from flanking the left of the troops that were [already] engaged. The regiment, and one from Minnesota, led by Gen. Heintzelman, obeyed the command with alacrity, and double-quicked through the fields, and Bull Run, which was three feet in depth and twenty yards in width at this ford. The water was yellow with mud, and flowed between the banks of red earth that showed the abundance of the sandstone in the soil. The soldiers followed the road over which the foe had been compelled to retire,[16] and deployed in line upon the ground on which the rebel battery, which opened the contest, had been planted. The strange spectacle of dead and wounded men scattered upon the battle-field affected all with peculiar sensations. While the regiment was moving to the front, Generals Heintzelman and McDowell, pointing in the direction of the firing, exclaimed, "They are running! The day is ours! They are on the retreat!" and one of them remarked, "Men, I pledge you my word of honor that there are not three hundred rebels on that hill."[17] When they reached the scene of contest, many were in that state of fatigue in which it was more natural to sleep than to fight. The regiment was shielded from the fire of the enemy at this time by the crest of a hill, upon the slope of which it was posted. The batteries of Griffin and Ricketts,[18] planted in a field upon the right, were actively engaged; and shell and solid shot were thrown with rapidity. The attention was excited by the singular shrill whistling that accompanied the passage of the balls and bullets through the

---

16 Outnumbered by the divisions of Hunter and Heintzelman, and outflanked from the east by Sherman's Brigade (Tyler's division), the Confederates retreated from Matthews Hill.

17 Gen. McDowell was indicating Buck's Hill, the next hill south of Matthews Hill, just north of the Warrenton Turnpike. [Memoirs]

18 Battery D, 5th U.S. and Battery I, 1st U.S.

air; but no symptoms of general uneasiness or fear were displayed. The line advanced to the crest of the hill [Buck's], and saw the enemies of the country.

They had been forced to quit the height near the Henry House; and the remnant, about fifty in number, was running in great disorder, and entering the woods, when the regiment delivered its volley, and many soldiers, like the author, discharged the first bullet from their muskets.[19] The foe was concealed in heavy force upon the left, and quickly returned the fire, when the order was issued to "lie down and load again;" and the smoke of rifles held by invisible hands formed the next target. The hostile batteries were masked in ravines and dense thickets; and white, sulphurous clouds, rising slowly at certain points, and the reports which constantly greeted the ear, were the only indications of their presence. While the officers were re-forming the lines, which were sometimes disarranged in the excitement that prevailed, Gen. McDowell and some members of his staff, together with other officers that composed a group of twelve or thirteen persons, rode to this position, and reconnoitred the woods and hills in front.[20] The soldiers were surprised to witness the boldness that was thus displayed; and expected to see them fall, but were amazed when they retired without receiving a bullet. When the regiment, inferring that the rebels had been forced from their last line of defense, advanced to the same point,

---

19 Franklin described the scene: "While [Ricketts] battery was in its first position, the Fifth and Eleventh Massachusetts Regiments were brought to the field, and took position just behind the crest of a hill about the center of the position. Here they were slightly exposed to the fire of the enemy's battery on the left, and were consequently thrown into some confusion. They fired without command, and in one or two instances, while formed in column, closed in mass." [OR II, 406]

20 From his position on Buck's Hill, Blake observed McDowell and his staff reconnoitering Henry House Hill.

a shower of lead welcomed it, and traced little channels in the ground upon which it stood. The enemy was again pushed back: the brigade filed to the right, and held a portion of the Leesburg Road, which ran through the stream at Sedley's Ford; and victory seemed to be no longer a matter of doubt. The national troops had pursued a retreating army a mile and a half: the Warrenton Turnpike was in their possession; and the left wing of Beauregard's force had been completely turned, so that his line of battle was formed at right angles with the Bull Run, which they had vainly attempted to defend. A lull in the infantry firing[21] took place a few minutes after one o'clock, and continued half of an hour; during which the men that were unoccupied should have attacked the rebels, who were enabled by this blundering delay to re-organize their shattered ranks, and offer a firm resistance when the offensive was assumed. These were the precious moments when a small fraction of the large reserve[22] should have been ordered to complete the triumph that had been already won. The commanding general never submitted an excuse for this omission of duty that satisfied those who took an active part in the engagement. I had a good opportunity to notice the topography of this portion of the field, which became the centre of the most stubborn fighting, while the soldiers were waiting for orders in the narrow Leesburg Road, that had been excavated seven or ten feet below the surface of the adjoining ground, and made a fair protection against an assault. There was a small wooden house, occupied by an infirm old lady, Mrs. Henry, who languished upon the bed of sickness during the contest, and was

---

21 Union forces advanced to the Warrenton Turnpike and paused at the base of Henry House Hill to reform. Meanwhile, Jackson's Confederate brigade arrived on Henry House Hill.

22 Two fresh brigades of Tyler's division were available just east of Bull Run. Beyond that, the nearest Union reserves were at Centreville, about four miles away.

killed by the troops who fired at the dwelling when it was filled with rebel sharpshooters.[23] The open space of ground was very irregular, and located between successive chains of abrupt hills about a quarter of a mile apart, which varied from one to two hundred feet in height. Thickets of pine and oak flourished upon the parts of the field in which the lines of the enemy were established; and the country was adapted by nature for defensive purposes, so that the rebels, when dislodged from one stronghold, always found another a short distance in the rear of it. They were concealed in forests which no telescope could penetrate; but the formation of the Union divisions took place in the open ground, and could be easily perceived by the hostile generals from the summit of a commanding height that overlooked the scene. The principal portion of the cleared soil was uncultivated, and covered with dry grass, and the black weeds which thrive upon land that has been poisoned by the culture of tobacco.[24] The rain had formed brooks that cut numerous deep gullies[25] in the slopes of the hills and every section of the field, and afforded an excellent refuge for the wounded who could not be carried to the hospitals.

The brigade was near the centre of the line at this time, and missiles of lead and iron were continually flying over it. Although the batteries had been placed in the extreme advance,[26] contrary to the well-known precepts of military authors, the success which had hitherto followed our arms tempted Gen. McDowell to make greater risks. The regular artillery, which had

---

23 Mrs. Henry's house was left in shambles after the battle.

24 Blake was a life-long non-smoker.

25 There had been no recent rain before the battle; Blake apparently is speaking of how rain had shaped the landscape over time.

26 Referring to Griffin and Ricketts used successively on Matthews Hill and Buck's Hill.

rendered splendid service, was removed from a secure position, and pushed to the open field in front, which was destitute of any natural barrier or protection.

"Put the battery upon that hill, and the day is ours!" shouted an officer of high rank.[27] The order was promptly obeyed, and the nation lost the victory. Members of the regiment destroyed a portion of the Virginia rail-fence upon the sides of the road, to allow the cannons to be drawn to the new point that had been designated. The gunners were proud of the success which had been achieved; and one of them said with truth, as he rode by the company, "We made the secesh battery change position three times in half an hour." The regiment was not actively engaged at that moment; and most of the men were watching the section of Griffin's Battery, which had been planted near them.[28] A heavy [Confederate] volley from many rifles ended the silence that had existed in the infantry-firing, before it had discharged two rounds: horses, officers, and men were killed or disabled in the space of a second; and, during my subsequent experience in a score of engagements, I never saw the work of destruction more sudden or complete. The battery of Ricketts, which was in line of that of Griffin, had been annihilated in the same decisive manner before the support[29] could be placed in position. I did not satisfactorily ascertain by whose stupidity this body of rebels was permitted to approach within two hundred feet of the lines

---

27 McDowell continued to push Ricketts and Griffin to the front – this time to Henry House Hill. Griffin protested, but he was promised support, so the two batteries trudged up the hill and unlimbered within 300 yards of the Confederate line.

28 Initially, Griffin deployed his entire battery on the left of Ricketts, but within half an hour, he was ordered to move two guns to the right of Ricketts. At this point in the battle the 11th Massachusetts was behind Griffin's initial position on Henry House Hill.

29 11th and 14th New York, plus a battalion of Marines.

without molestation,[30] until I read the testimony of Gen. Griffin before the Committee on the Conduct of the War. I quote his language, because this disaster was the first check that had occurred in the action, and the chief cause of the defeat. "Major Barry said, 'I know it is the battery support: it is the regiment taken there by Colonel _____.' I said, 'They are confederates; as certain as the world, they are confederates.' He replied, 'I know they are your battery support.'"[31]

The few cannoneers that survived this fatal volley immediately rushed to the rear: wounded horses, in their agony, galloped through the ranks of the infantry, and trampled upon the dead and helpless who were lying upon the field. Three animals, which were harnessed, and attached to a caisson, dashed through the regiment at a furious rate of speed, and dragged one that was severely injured: a soldier, whose legs had been shattered by a solid shot, sat upon the carriage, clinging to it with his hands; and a stream of blood sprinkled the earth, and made a trail by which the course of the caisson could be traced. The troops now lost the confidence of victory which they had hitherto possessed; while the defeated and disheartened rebels, who saw eleven pieces of [Union] artillery in an instant placed *hors de combat*, at once renewed their efforts, and their yells of exultation were heard above the din of the conflict. A squadron of their cavalry attempted to make a charge; but many of their saddles were

---

30 This was the 33rd Virginia, which had avoided "molestation" because they were clad in blue. Uniform confusion was common during the battle. The 11th Massachusetts wore gray uniforms "which had been sent us by the state committee and which we were forced to pay for, despite our protests. The delusive color cost us dearly, for, at the first battle of Bull Run, one of our regiments opened fire upon us, mistaking us for Confederates, and several valuable lives were sacrificed." [Hutchinson, 22]

31 Maj. William F. Barry was Chief of Artillery. The unnamed colonel was Heintzelman, personally directing the 14th New York. [OR II, 347]

emptied, and they were easily repulsed by a body of men who belonged to different regiments.[32] In the mean time, the rebel leaders had rallied their stragglers and fugitives, and advanced their lines to capture the guns which were now powerless to do them any injury. They were triumphantly driven three times to the woods, and victory was once more within the grasp of the Union general.[33] The soldiers seized some of the pieces, and pulled them a few yards to the rear; but were compelled to leave them, because the defeat of the enemy required the presence of every man in the line of battle.[34] The regiment at one time stood upon the ground which was held by the foe when the first volley [had been] discharged, and the dead and wounded were seen upon every side. The Mississippi troops[35] had sustained a heavy loss at this point; and one of them, who was dying, remarked to the men, "You have fought for your country, and I have fought for mine." The mangled artillerists rested beneath the guns, in serving which they had so bravely fallen.

It was three o'clock, and the soldiers had been engaged upon the march, or in action, during the long period of thirteen hours. A large number, from various causes, had left their commands

---

32 Stuart's 1st Virginia Cavalry was driven off primarily by the 11th and 14th New York. [Ibid.]

33 "Entering the field on the right of the battery [Ricketts], we halted our right, while as fast as the regiment passed in it was thrown forward, left, into line, clearing the ground from the rebels and re-capturing the guns. The regiment now occupied a line nearly parallel to the turnpike, reaching from the top of the hill to the valley near Young's Branch." [Hutchinson, 24]

34 Franklin ordered the 5th and 11th Massachusetts to "try to get back the guns. They went there, and with the help of some other regiments on their right, the enemy was driven from the guns three times. It was impossible, however, to get the men to draw off the guns, and when one or two attempts were made, we were driven off by the appearance of the enemy in large force with heavy and well-aimed volleys of musketry." [OR II, 406]

35 Jackson's brigade included the 2nd and 11th Mississippi.

and escaped to the rear, or fought without regard to the rules of discipline; but the colors of the regiments, and the organizations, with many of the officers and companies, still remained. The exhausting march, the terrible heat, the lack of water, the horrors of the battle, and above all, the loss of the artillery, had affected those who remained, to such an extent that they became every minute more unfitted to resist the onset of the enemy, who maintained an irregular fire from the forest. Some officers behaved in the most cowardly manner; and certain companies were commanded by sergeants, because the captains and lieutenants absented themselves during the engagement.[36] An uninjured colonel, who pretended to be severely wounded, and declared that he was unable to walk, was borne from the field by four members of his regiment.[37] There was no general demoralization in the army, although many of the troops acted like novices in the dreadful art of war, and executed some movements with great confusion.[38] Two men placed their hands upon their ears to exclude the noise of the musketry and artillery, and rushed to the woods in the rear of the regiment. A timid

----

36 Including Company D: "The officers of his company either abandoned the men or were driven off, when [Gustavus] Hutchinson, who was Orderly Sergeant, acted as Lieutenant, and Sergeant [William] Joy as Captain, with thirty men, made several charges, and finally left the field in good order." *Boston Daily Courier*, August 1, 1861.

37 Possibly Col. Clark – "Doubtless you have heard of Col. Clark, at the Bull Run battle; but I won't say anything about him, as you have seen in the Boston papers as much as we know about him; but you must not think that his men acted in the same manner. They walked from 11 o'clock Saturday night before the battle, until 10 o'clock Sunday morning, and then went into that fight without even an hour's rest or anything to eat, and fought bravely." *Cambridge Chronicle*, August 24, 1861.

38 Capt. D. P. Woodbury, of Heintzelman's staff, noted the situation around 4 P.M. "The instinct of discipline which keeps every man in his place had not been acquired. We cannot suppose that the troops of the enemy had attained a higher degree of discipline than our own, but they acted on the defensive, and were not equally exposed to disorganization." [OR II, 334]

Catholic took his service-book from his pocket, and read some of the prayers when the brigade was posted in the road. The shells struck rifles with such force, that some were twisted into the form of circles. A cannon-ball severed the arm of a sergeant, and threw it into the face of a soldier, who supposed, from the blow and the amount of blood upon his person, that he was dangerously wounded. One man stumbled over some briars while the column was ascending a hill; when a solid shot passed over him and killed his file-leader, when he fell upon the ground. The ghastly faces of dead, and the sufferings of the wounded, who were begging for water, or imploring aid to be carried to the hospital, moved the hearts of men who had not by long experience become callous to the sight of human agony.

The firing in front was very feeble at four o'clock; but a succession of severe volleys[39] was poured from the woods upon the right flank and rear. The troops were unable to offer any resistance, and began to retire from the field upon which they had maintained, unaided, the long struggle. The fresh [rebel] soldiers that arrived at this opportune moment belonged to Kirby Smith's brigade, and formed a part of the rebel army of Johnston in the Shenandoah Valley. Generals and mounted officers, among whom were Burnside, Wadsworth, and Gov. Sprague,[40] attempted

---

39 Pvt. Edward Gazet of the 11th described what it was like to be under fire: "When I look back and think of the scene it seems I bore a charmed life for I kept in the open field and was in the most exposed situations, at one time I went over to a side of the field and opened a haversack and took a biscuit and began to eat it, when all of a sudden there came a volley out of the woods a few rods from me and two balls passed one on each side of my face so near that the wind of them made my ears smart and one struck between my feet and one on each side of me. I stood still a moment and thought it would be best to clear out or lie down ... I decided on the latter and did so and munched on my crackers cheered by the whistling of bullets and the occasional bursting of a shell." [Gazet letter, July 23, 1861]

40 Burnside led a brigade of Rhode Island regiments; Wadsworth was on McDowell's staff; Gov. William Sprague of Rhode Island accompanied his

in every way to form a new line, and prevent the retreating regiments from leaving the field; but the position that was selected had no natural strength, and could be discerned by the enemy, who followed with the energy of conquerors. The colors of some commands were planted firmly; and every man was ordered to rally around them, and make one more effort to win the battle; and officers shouted, "Rally round the old flag!" or, "Zouaves, remember Ellsworth!"[41]

When the foe advanced with loud yells, and it was certain that the thousands who had remained in the reserve at Centreville, and rendered no service during the protracted contest, would not assist the diminished numbers that were formed upon the field, despair was visible on every face; and the regiments fell back about half-past four o'clock. The rebel artillery opened as I passed the Stone Church,[42] which had been used as a hospital; and their cavalry followed at a safe distance in the rear of the mass of the army, after they had received a few rounds of cannister. The men were overcome by their thirst; when they forded the run, and drank copious quantities of the water, which was constantly disturbed and filled with particles of mud by the tramp of horses and soldiers through it. The disorder that existed was increased at this point; and the sorrowful troops, who had been forced to show their backs to the enemy, discussed the causes of the repulse while the shells were bursting in their midst. All seemed to wish to reach some rallying-point like Centreville, so that they would not be taken prisoners; and some

state's troops but did not hold army rank.

41 Elmer Ellsworth had raised a regiment from the firemen of New York City (11th New York *Fire Zouaves*) with uniforms styled on French colonial soldiers of Algeria. Ellsworth was shot dead two months previous to Bull Run while removing a Confederate flag from the roof of the Marshall House hotel in Alexandria.

42 The stone church was Centreville Methodist church, constructed in 1854.

officers removed their shoulder-straps to conceal their rank, and rushed to Washington. The [rebel] infantry did not attempt to pursue the retreating columns; and the cavalry halted for the night upon the south bank of Cub Run, which flows into Bull Run. The disorganized brigades marched upon the road over which they had passed in the morning; squads were scattered in all directions, stopping occasionally to eat the refreshing blackberries under their feet; and few bodies of men were moving with regularity.

The gunners of the Washington Artillery [of New Orleans] obtained an excellent range upon the bridge over Cub Run, and demolished an army wagon, which was not removed by the teamsters who blocked up the way in their eagerness to escape. The stream was not fordable: trains and batteries that had not yet crossed were abandoned, and one-third of the loss that was sustained in the munitions of war occurred at this place. The general that allowed the wagons to go to the front committed an inexcusable error: if they had been parked at Centreville with the reserve, not one of them would have been captured. The so-called panic, about which so much has been said by persons that have given a description of the battle, occurred at this time. The drivers, finding it impossible to cross the run with the wagons and artillery, took their horses, and sometimes cut the traces to expedite their movements: because the shells were continually bursting near them, and there were no troops upon that side of the stream to resist the [rebel] cavalry if they made a charge. Some government teamsters, who belonged to no army organization, and were upon the safe bank, beyond the range of the rebel guns, cowardly deserted their wagons, and rode, without halting, until they saw the dome of the Capitol. The foot-soldiers, alarmed by this strange conduct and the absence of general officers, double-quicked and run; and hundreds cast aside muskets, axes, and equipment, so that their flight could not be retarded. The appearance of so many full regiments at Centreville, that had been unemployed during the day, caused much excitement; and the troops that had undergone the perils of the fight were very severe in their comments upon the ability and loyalty of the commanding general. "We have been sold," was a common remark in their conversation. The last shot was fired a few minutes before sunset; and the armies no longer heard "in tones of thunder, the diapason of the cannonade."[43] The shells and bullets ceased to sing their songs of death in the forests of Manassas, but rushed in silence, until they struck the homes of their victims,[44] in the peaceful villages of the north.

I was obliged to leave the ranks during the latter part of the march, on account of exhaustion produced, in the battle, by a fragment of shell which had inflicted a mere scratch.[45] I walked in the direction of Centreville, at daybreak, on the 22d, after a sound sleep in a clump of bushes, and expected to find the army established upon the heights. In traveling on the road which led through this town and Fairfax Court House, the amount of

---

43 *The Arsenal at Springfield* (1845) by Longfellow.

44 11th Massachusetts lost 15 killed, 43 wounded, 30 missing. [Hutchinson, 25] Colonel Clark stated that the 11th "carried upwards of 800 men into the action." *Boston Herald*, July 30, 1861

45 Blake described his wounding: "When the 11th was posted in a sunken road near the Henry house and facing south, a rebel battery was active and had a good range on our position. A shell exploded, a fragment breaking the foot of Joseph Kent, a private in the company, and I was struck in the left breast by a hard substance packed inside. It cut through my blouse and undergarments and scratched my skin and hit with such force I was weak and could not stand erect. Nearly two hours elapsed before I recovered my wind and in the meantime, I was in a gully washed by heavy rains in the red soil, and at one time was between both armies, and bullets were whistling to and from over me. Fortunately, the enemy was driven back and I was in the ranks of the 11th when it retired from the field." [Memoirs] Joseph Kent was discharged September 26, 1861. [MSSM/1, 811]

government property that was needlessly destroyed, during the retreat, was easily ascertained. There were ten or twelve commissary and ammunition wagons in the streets of Centreville; and three had been abandoned within a mile of the last-named place, when the rebels were at least ten miles from them. Blankets, rifles, and equipments of many descriptions, were scattered in the road, and the woods that bordered upon it; and some had been thrown away by men who were near Alexandria. Crowds of women and negroes, like wreckers in a stranded ship, were taking flour and provisions from the deserted wagons; and the commissary department of the enemy obtained a small number of rations at the expense of the national treasury.

The houses upon the public way, and especially those of Centreville, were filled with the wounded who could not walk: there were no surgeons or nurses to dress and bandage their injuries; and they implored all the able-bodied persons to tell the general to send doctors and ambulances. Squads of stragglers, and slightly wounded men, with bandaged heads, arms in slings, and limping upon sticks, were walking to overtake the army, which had marched during the night. A steady rain fell during the day; and with my musket, and equipments for companions, I arrived at night at Alexandria,[46] completely saturated. Upon the line of retreat, the natives, comprising old men and the female portion of the community, openly expressed their joy at the result of the conflict, and misled the soldiers by willfully deceiving them about the direction of the roads; while others, and the Irish settlers near the railroad, in every way assisted the stragglers. When I passed through Centreville at half-past seven o'clock in

---

46 The 11th "retreated to the ground it had left in the morning, east of
   Centreville, and remained there until ordered to its old camp at Shuter's
   Hill, which we reached the following day (Monday) utterly fatigued by our
   continued exertions. Our regiment brought back its colors, four banners
   having been carried in the fight." [Hutchinson, 24]

the morning, a loyal man, without any hat upon his head, which was adorned with the white hair of age, stood at the intersection of the streets, pointed out to all that which led to Fairfax Court House, and earnestly advised them to hurry as much as possible, because the rebel cavalry could cut them off as soon as they knew that the troops had marched to Alexandria. Most of the wagons which had been purposely or shamefully abandoned were marked "U.S." and I did not see a regimental team upon the route. There was an important distinction between the drivers: those of the first were hireling civilians, while the last were soldiers detailed for this duty from the volunteers.[47]

Near Fairfax there was a squad of fifty men, two of whom had fastened white handkerchiefs to their bayonets to prevent the rebels from firing upon them. Other groups marched together under a commander of their choice, kept good order, and avowed an intention to resist the cavalry, which was momentarily expected. Three men tied their muskets to the saddle of a horse which they had found, and each one rode a portion of the way. A soldier knocked down an officer who was mounting his steed, jumped upon the animal, and, in a few minutes, there was an impassable gulf between the owner and the thief. Many of the ambulances and wagons, from which the stores had been removed, instead of conveying the wounded, were crowded with officers and men who wished to secure a ride.

The reports of the fighting at Bull Run were distorted accounts of a single feature of the retreat;[48] and the journals and people spoke of nothing except the "panic-struck troops" or a

---

47 Blake is making an esoteric point about the teamsters of the supply wagons: civilian contractors in the Federal army vs. enlisted soldiers in the state regiments.

48 The newspapers emphasized the Washington Artillery shelling of the Cub Run crossing and the resulting confusion.

"routed army." Certain generals and staff officers shrewdly and dishonorably availed themselves of this fact, and threw the cause of the defeat upon the "disorganized volunteers," to shield themselves from the share of public censure which they justly deserved. Major Barry, with remarkable assurance, testified concerning "uninstructed," "raw," and "green" troops, "panics," "indolent officers," and "infantry support broken in confusion, and scattered in all directions." The wisdom of Gen. Scott,[49] in opposing the appointment of Gen. McDowell, was fully confirmed; and the soldiers that formed his command considered that his incompetence was the primary reason of the unfortunate defeat. The effective strength of the army was diminished by the mode in which it was maneuvered and separated. Gen. Runyon[50] was stationed near Fairfax Court House, which was a day's march from the battle-field, with seven or ten thousand men, and performed no more service for the country than the Queen's Guard in London. Another body of eleven thousand troops, under Col. Miles,[51] who was intoxicated, and unrelieved when the fact was reported at headquarters, was posted in the morning at Centreville, upon the left, and remained there during the day, without rendering any aid to their comrades upon the right. A few skirmishers rarely exchanged shots; and the artillery quickly dispersed a small rebel force that reconnoitred the position by

49 Winfield Scott, septuagenarian commander of the Union army. In fairness, McDowell did express concern that the Union army was too green, to which Lincoln replied that both armies were "all green alike." McDowell was under pressure to act before the 90-day regiments went home. Scott retired in November 1861, succeeded by Maj. Gen. George B. McClellan.

50 Mayor of Newark, Theodore Runyon, was appointed to command the 4th Division of mainly New Jersey regiments that were the greenest of the green.

51 Dixon Miles, 5th Division commander, was accused of drunkenness at Bull Run. He died the following year at the siege of Harper's Ferry, which resulted in the surrender of its 12,000-man garrison.

firing nine or ten rounds of cannister. Nearly two brigades of the division commanded by Col. Tyler were halted upon the north bank of Bull Run, and most unwillingly acted the part of spectators. It will be observed that less than three-eighths of Gen. McDowell's force (about fifteen thousand men) actively participated in the combat; the remainder (about twenty-five thousand) did not fire a cartridge at the enemy. Of forty-nine pieces of artillery which was attached to the army, only twenty-two were planted upon the field of strife. This small number of gallant soldiers, at times basely deserted by certain brigadiers, overcame serious obstacles, gained a brilliant success, which was not followed up, and was finally repulsed by fresh troops. If the list of casualties is apportioned among those that actually fought, and it is remembered that the contending troops were unused to the disturbing events of battle, and could not aim with the deadly accuracy of veterans, their bravery is vindicated. It is an interesting fact, that, while the so-called generals of the rebel army suffered a severe loss,[52] not one was killed upon the Union side.

Many negroes gladly escaped from the lines of the enemy, and brought valuable information; but their statements were unnoticed: and, contrary to every dictate of humanity, they were forced to return to their masters, and crouch in helpless agony beneath the cruel lash of fiends. Thus Gen. McDowell blindly rejected the best means of learning important facts, and gave to the foe some laborers, who worked upon the forts of Centreville and Manassas.

The reports of rebel generals and authors, their maps, and especially that which was "taken by Capt. Samuel P. Mitchell,

---

52 Confederate officers killed in action included Brig. Gen. Barnard Bee, Col. F. S. Bartow, and one regimental commander.

First Virginia Regiment,"[53] who took part in the battle, concur with Jefferson Davis in stating that it was "a hard-fought field."[54] Prisoners always admitted that the arrival of Smith's brigade saved them from utter defeat. Every general and civilian that prefers charges of cowardice,[55] or "panic-struck" flight, against the troops that attacked the foes of the country in the first action of Bull Run, is a base slanderer or a culpable ignoramus.

---

53 Samuel Mitchell's hand-drawn sketch map of the battle uses the words "fierce fighting here" several times in its explanatory notes.

54 Letter from Jefferson Davis to the CSA Adjutant-General, July 21, 1861. [OR II, 987]

55 Pvt. Timothy Emerton wrote his sister: "I suppose you read of the battle of Bull Run on Sunday the 21st of July we was in the thickest of the fight yet the papers do not give us any credit we are no cowards if we are green at the business but next time look out for a different story for their ill-using our wounded." [Emerton papers. USAMHI]

Chapter II

The Camps at Bladensburg and Budd's Ferry

THE DEFEAT at Bull Run disheartened the troops; and, like most soldiers who have taken part in an engagement, they did not wish to witness another [defeat] if it could be honorably avoided. Many who had approached the field with misgivings about their courage in the presence of death were happy in the thought that they had performed their duty without displaying any symptom of fear. The regiment returned to its old quarters at Alexandria; and, with the exception of frequent night-alarms concerning the rumored advance of Johnston and Beauregard to attack Washington, nothing of interest occurred.[1] The stores of the city were closed, a regiment composed of its citizens was in the rebel army, grass was growing in the main streets, and a perpetual sabbath reigned. Every soldier visited the Marshall House; and, at the time I saw it, the stairs and doors had been wholly cut away by the hatchets and knives of those who carried away some relic to "remember Ellsworth." The hostile opinions which the people entertained towards the Union forces was expressed by a daughter of a wealthy traitor, who remarked to the sentinel at the principal entrance to the camp, "Only niggers and trash come to see you."

---

1 Senator Henry Wilson visited the camp named in his honor and he "suggested he could help the boys by the use of his franking privilege, and it was arranged that I should write his name on the mail matter of the 11th. He wrote his official signature on a sheet of paper and handed it to me, and I imitated the same as accurately as I could for four months, [until] the Congressmen who had done likewise discontinued the practice because the postal authorities complained and made reports against it. I was appointed Regimental Postmaster and, while the 11th was encamped at Alexandria and Bladensburg, was furnished with a government horse branded U.S. and rode daily to the post-office in Washington." [Memoirs]

The regiment marched upon August 9, and pitched its tents upon the battlefield of Bladensburg during the following day.[2] The famous dueling ground[3] was within the lines, and furnished an excellent place for target practice. The First and Eleventh Massachusetts Volunteers, Second New Hampshire Volunteers, and Twenty-sixth Pennsylvania Volunteers, formed a brigade, and Gen. Hooker was assigned to command. Orders relating to the discipline of the troops were issued; and the hours for eating, drilling, and sleeping, which comprise the chief duties of military life in camp, were indicated by the bugle and drum. The *reveille*, which, like the voice in the colossal statue of Memnon,[4] welcomed the rising sun, was the signal for the soldiers to "fall in for roll-call;" and those who delayed to obey the notes of the drum and fife were seized by the strong arm of "brief authority,"[5] in the person of a sergeant or corporal, and brought into the company street. Gen. Hooker always inspected the brigade at this early hour, and summoned officers that were absent, or sleeping in their tents, to drill their commands. Seven hours were daily occupied in drilling, and one that preceded "breakfast at seven" was the first that claimed the attention. Unlike Generals McDowell and Franklin, who always exhibited towards subalterns, and especially enlisted men, the most supercilious bearing, the brigadier listened to every person — the drummer-

---

2   Americans troops were soundly defeated at Bladensburg, August 24, 1814, leading to the British burning of Washington.

3   From 1808 to 1868, a clearing in Bladensburg was the site for dueling with pistols and sometimes even with muskets. Among the dead was naval hero Stephen Decatur (1820). For an account of the dueling history of Bladensburg see Cudworth.

4   Memnon in Egypt has been legendary since antiquity. It was said that when the first rays of the dawning sun fell on the colossi, they would hum like a plucked harp-string. [Hamilton, 290]

5   *Measure for Measure*, Act 2, Scene ii (Shakespeare).

boy or the colonel of the regiment — with such candor and sincerity that he quickly acquired their confidence.

The negroes that lived within fifteen miles of the camp walked to it upon Sunday, and brought small quantities of fruit, which was generally carried upon their heads. Although their masters claimed that they were well fed, all the unconsumed food was given to them by the soldiers, who watched them with amazement while they devoured enormous quantities. They collected the clothing that had been cast aside, and wore the uniforms, but since that date established their right to them by enlisting in the service.[6] They were deceived by the whites in regard to the treatment they would receive; and the sight of a musket in the hands of one of the men caused general terror, and earnest prayers to spare their lives. They imagined that their shackles would soon be broken, and manifested their joy by queer songs and frantic dances.

The command was sometimes under orders to move at a second's notice; rations for three days were cooked, and the cars stood upon the railroad to receive them. The regiment performed picket duty in the town, and searched every wagon that passed over the road to Washington to find contraband mails and ammunition; but a few peaches and water-melons were the only articles that were confiscated. The heat was so intolerable during the day, that all clothing seemed to be superfluous; but the cold chills in the night would be so intense, that overcoats would be required to keep their wearers comfortable. Necessity compelled the men to learn many domestic duties. The little streams near the camps were always occupied by groups who were washing their clothes or persons; and the brook, besides its value on account of the water, furnished a mirror; and the bushes that

---

6   Blake refers the U.S. Colored Troops, which by the time he wrote his book, had enlisted around 175,000 men.

grew upon the banks were covered with towels and clothing. The bed was easily made by placing a rubber blanket upon the ground, and using one of wool, and an overcoat for quilts; while a pair of shoes and a knapsack formed the pillow. The ventilation of the tents, it is needless to say, was perfect; and the effect of simple food, exercise, and pure air, upon the health of the troops, was excellent; and some added thirty pounds of flesh to their bodies during the first three months of their service. Although many officers exposed the frauds of government contractors and inspectors, the army was bountifully supplied with defective articles of every description. The tents leaked "like sieves," until the general [Hooker] succeeded in procuring those of another pattern, that were useful when it rained. Many pants and blouses, which had been worn with ordinary care, were reduced, at the end of two weeks, to worthless rags, that no "stitch in time" or sewing machine could unite together. The soles of thousands of shoes, if once partially wet, had the qualities of pasteboard; and some blankets were as valuable for comfort as those which Surgeon-Gen. Hammond purchased for the hospitals.[7] The slight experience of a month satisfied the troops that the patent drinking tubes and filters, which adorned their necks when they marched through the cities *en route* to Washington, were of little if any use; and they were cast aside. Havelocks, which, besides other

---

7 Blake sarcastically refers to William A. Hammond, who proposed several significant reforms to the medical department, many of which were not adopted until decades later. "On account of charges preferred against him of irregularity in the award of liquor contracts, Hammond was tried by court-martial and dismissed from the army in August 1864, but in 1879, upon a review of the court-martial proceedings made by the president, he was restored to his place on the army rolls as surgeon-general" [Union Army, vol. 8] Blake's own experiences with the medical department (Chapter XVIII) possibly soured his opinion of Hammond. Then again, Blake, a strict temperance advocate, may have considered the problems relating to the liquor contracts as evidence of corruption.

inconveniences, did not avert the rays of the sun, were torn into shreds, and furnished rags for cleaning guns and swords.[8]

Earthworks were constructed around Washington; and details were daily ordered to dig trenches, and fell the woods and orchards that interfered with the range of the redoubts and forts. Drills, fatigue-duty, and a review by the President and Gen. McClellan, were the only events that disturbed the usual quietness of the brigade camp.[9] The command of Gen. Sickles[10] and that of Gen. Hooker constituted, in the month of October, a division which was placed in charge of the latter, and broke up its encampment upon the 24th. I heard a conversation, and observed certain facts in one regiment of the brigade previous to the march, and narrate them to show the manner in which some chaplains discharged their religious tasks. Two hundred Bibles and Testaments, and a package of tracts, had been sent to this officer for distribution; but his time had been occupied in acting as purveyor and cook for the field and staff, and he had neglected to circulate them. When the marching orders were received, he was very industrious, and carefully packed his large stock of crockery, and private property, including a faded, good-for-nothing umbrella, for transportation, and sold the floor of his

---

8   The "drinking tube" filters fit into the spout of a canteen. "Havelock" was a white linen or cotton "apron" worn over the cap so as to shade the back of a soldier's neck from the sun. Named for Sir Henry Havelock, whose troops in India had used them in the 1850's. Women of the north seized upon the making of havelocks as a favorite method of supporting the troops.

9   Blake was occasionally detailed travel to Washington for a supply of bread "which was baked in ovens in the basement of the Capitol. It is difficult to believe today [1916] that the National Capital and its suburbs had at this period poor and narrow streets and that I saw hogs wallowing in the mud holes of Pennsylvania Ave." [Montana, 35]

10 Sickles raised four regiments of the Excelsior Brigade and was given command of it even though his prior military rank had only been as a major in the New York militia.

tent, consisting of eight or ten boards, to a citizen. The colonel said to the chaplain, when the column began to move, "Have you sent all your baggage to the train?" — "Yes," he replied; "but what shall I do with these things?" — "What things?" — "Nothing but some religious matter," he answered, as he pointed to the Bibles and tracts which had been scattered upon the ground by the purchaser of the boards. "Why don't you give them to the men?" — "They don't want them," the chaplain quietly remarked; and this gift of Christian friends was left upon the field. The same person was ordered to preach, three months after this occurrence; and obliged to borrow a Bible of a private in his regiment, because he did not have one copy in his possession.[11]

The division halted and encamped at a point near Budd's Ferry, upon the Potomac, after marching fifty-five miles in four days through a thinly settled section of Maryland that contained a few hovels which their inmates styled towns and villages. The troops bivouacked during the nights in the woods and fields near the road; and the blazing camp-fires, made chiefly of rails taken from well-seasoned fences, were the centres of circles twenty or thirty feet that belonged to the sleeping soldiers. The "general" was played in the morning, to notify the men to prepare for the march;[12] and the fifes spoke the words that the veteran musicians ascribed to them: — *Don't you hear the general say, 'Strike your tents, and march away'? Yes: I hear the general say, 'Strike your*

---

11 Blake had nothing positive to say about the chaplain of the 11th, Elisha F. Watson: "The love of the root of evil was the ruling passion of our Chaplain. [He] preached one sermon on Faith and another on Hope, and then reported his valise and fifty sermons were stolen and the contents were never recovered. He said they were the products of twenty years study and labor and did not preach again to the 11th." [Memoirs]

12 The "general" in this context was a fife and drum call used to assemble all hands in readiness for a move.

tents, and march away' – The roads, like those of Virginia, were in a miserable state, and the art of building bridges was unknown. The rations consisted of salt pork, which was usually eaten without the aid of any culinary process, and the regular quantities of coffee and hard bread.

The division encamped at the distance of two miles from the Potomac; and the regiments daily furnished details that performed picket-duty on the banks of the river, and more than fifteen miles were carefully watched to prevent illicit trade between Virginia and Maryland. The cold winds might penetrate the frame, snow and sleet might chill the limbs; but the vigilance of the Argus-eyed[13] sentinels never ceased. The rebels had constructed breastworks and forts in the woods, upon the steep bluffs of the southern shore, on Cockpit Point and Shipping Point, near Evansport,[14] where the main channel ran near this bank of the river; and a blockade for the purposes of general navigation was established.[15] Before the trees which concealed the batteries had been felled, and while the captains were ignorant of the presence of an enemy, the hostile gunners seized two schooners and the steamboat "George Page,"[16] and anchored

---

13 Argus: a vigilant, multi-eyed giant of Greek mythology. [Hamilton]

14 Hooker's headquarters were six miles northeast of Budd's Ferry on Chicamoxen Creek. The brigade camped directly across the Potomac from Evansport (now Quantico Marine Base). The Excelsior Brigade camped to the south of Hooker's Brigade, with headquarters near Liverpool Point, extending the Union presence to Smith's Point opposite Aquia Creek.

15 Confederates batteries inhibited the passage of Union shipping on the lower Potomac. Besides significant political embarrassment, there were concerns that the rebels might cross the lower Potomac and stir up insurrection. To counter this possibility, Hooker's division was sent to Budd's Ferry. For an overview of the blockade and its effects on Union morale and military planning, see Wills.

16 CSS *George Page,* an ex-U.S. Navy transport, was fitted out for river defense; later scuttled when Confederate forces abandoned the Potomac position.

them in Quantico Creek. The regiment guarded the ground which was in front of the earthworks, and within the range of the cannons; and once in ten days the company, for twenty-four hours, patrolled the northern bank. The traitors' flag of three bars waved over the forts for three weeks; when it was lowered, and never elevated again. A large telescope mounted upon a tripod could be seen; but it was soon removed on account of the dangerous tendencies of the shells which the gunboats and a land battery threw into the work. Heavy siege guns, including some of English manufacturer, had been mounted in the forts; and from the embrasures, *Their mouths, with hideous orifice, Gaped on us wide.*[17]

The thirty-two pound Parrott gun, which opened the battle of Bull Run, and was lost at the Cub Run Bridge, had been planted in the lower battery, and was frequently fired by the rebels at a high elevation, until it burst, and killed and wounded more in its death than during its life.

Near a house[18] which had been recently occupied by its owner, an aged lady, Mrs. Budd, the men, under cover of night, threw up a small earthwork, in which two pieces of light artillery were placed. The foe, at first, often fired at the pickets without any serious results, and then attempted to destroy the little boats and smacks which sometimes sailed by. Three men were stationed, at certain intervals, upon every post near the river, in the vicinity of the hillocks and large trees which were essential to their protection and comfort; and the path that led from the right to the left of the line was torn, and many holes were excavated by

---

17 *Paradise Lost*, line 577 (Milton).

18 "Mrs. Budd had left her house…and it is now used by the officers and men of a section of artillery which is posted near it to warm their fingers and their coffee in. It is by degrees coming to pieces, the men stealing the clapboard and flooring to make themselves bunks." [Wainwright, 13]

huge shells from the guns in the "Old Dominion."[19] The guards dug eaves, and with cornstalks, boughs, and mud, constructed shelters and huts, which resembled in style the rude structures of savage tribes. The fire, fed by the driftwood gathered upon the shore, enlivened its warmth, day after day, and month after month, the pickets that in succession encircled and watched the cheerful flames like the vestal virgins of Rome, who nourished the embers that were never extinguished.[20] While the dark hours slowly sailed along, and seemed at times becalmed, thoughts of a cherished home arose in the mind of the soldier who — *Sat by his fire, and talked the night away.*[21] — When there was no breeze, a person could shout across the Potomac, which was a mile and a half in width at this point: the orders of officers drilling their commands, the sound of a solitary bass drum, and the strokes of axes, were distinctly heard. The hostile pickets sometimes conversed with each other, and profane taunts and coarse jokes concerning "blue-bellied Yankees," and "ragged and shoeless secesh," passed from one side to the other. If one of our cannons was discharged at the batteries, their sentinels immediately yelled, "Look out! the Yanks' shell is coming!" and similar outcries. The names of regiments and generals were ascertained in these conversations; and a rebel asked, upon one occasion, "Is that woman we can see riding upon a horse Gen. Sickles's wife?"[22]

Vessels loaded with cargoes of wood and hay, to the number of twenty, sometimes passed up the river during the night, and rarely attracted the notice of the [rebel] artillerists. The path of

---

19 Nickname for the state of Virginia.

20 Priestesses kept a fire burning perpetually in the Temple of Vesta.

21 *The Deserted Village* by Oliver Goldsmith (1770).

22 Apparently meant as a joke. Sickles' marriage was not a happy one subsequent to his gunning down his wife's lover in 1859. [Sassaman]

the shell through the air could be traced by the sparks that fell from the ignited fusee, and the deadly flames removed the obscurity of midnight when it exploded. Oyster-boats boldly glided along in the daytime; and immense amounts of ammunition were burned in the futile attempt to destroy them. It is very difficult for gunners to aim with accuracy at a moving object; and not a single vessel or person was seriously injured during the blockade, although at least eight thousand shots were hurled by the rebel cannons. The sailors concealed themselves in the hold; while the craft, urged by the breeze, moved on, and seemed to be guided by fairies. The batteries, hidden in clouds of smoke, were active until the cause that excited their fire was beyond their range; and the pickets on every part were successively exposed to the shots that passed over the boats. Balls and shells, varying in weight from twenty-four to one hundred and twenty-eight pounds, were often taken from the earth in which they had been buried. The enemy, in several instances, ascertained, by some inexplicable means, the countersign, and shouted it across the river before it had been communicated to the pickets by the brigade officer of the day. The action of some citizens who lived in Maryland, and were suspected on account of their well-known sympathies for friends in Virginia, were constantly observed. A singular waving light, which was moved from point to point in a dark and stormy night, was once seen, and the occupant of the house near the stream, towards which no gun was ever pointed by the rebel cannoneers, was promptly arrested and sent to Washington for the serious crime of making signals to the foe.[23] When the facts were

---

23 Capt. Robert S. Williamson (Topographical Engineer) mapped the area and reported that "Posey's house is on a hill about a mile back from the ferry, and his windows are in full view of the rebel batteries. He is undoubtedly in concert with the rebels, but tangible proof is wanting. It is believed, however, he communicated intelligence to the enemy by means of mirrors

investigated, it appeared that he had carried a lantern to some outbuildings to awaken cooks, and procure food for a guest who had arrived at a late hour; and he was released after he took the oath of allegiance.

The Southern winter differed from that of the North in its essential features: rain fell instead of snow, and the ground was covered with a thick layer of mud. Details, that occasionally comprised the entire force, labored and corduroyed the roads,[24] which were often impassable; and when the horses could not draw the necessary supplies in the army wagons, a barge, manned by soldiers of nautical experience, conveyed them in the night, within the range of the batteries, to a point near the regiment. No orders were received to go into winter quarters when the cold season advanced; and applications for leave of absence were generally refused at army headquarters in Washington, because the "exigencies of the service" required the presence of every officer and man with his command. The balloon[25] ascended to make reconnaissances; and it was supposed, from these facts, that the Government would grant Gen. Hooker's request to attack the batteries and re-open the Potomac.[26] In this state of uncertainty, many who had anticipated

---

and candles from his windows, the women of the house taking an active part in these proceedings. It is very certain the rebels know all that we are doing. If heavy guns are to be brought to the ferry, that house should be closed, or the inmates sent away." [OR V, 630]

24 A corduroy road had split logs laid side-by-side across the road-bed as paving, so that wagons would not sink into the soft ground underneath.

25 The balloon flights at Budd's Ferry were the first undertaken by Thaddeus Lowe for the Union army. Balloons were meant to be used at Bull Run, but were late leaving Washington. [Fishel, 37-38]

26 Hooker wrote to McClellan on February 22: "I would begin the attack on Cockpit Point, march down the river, crossing the Quantico by boats. With six...howitzers from high ground on the north side of the Quantico I can drive the rebels from the batteries at Shipping Point in two hours." [OR V,

a movement neglected to build comfortable barracks until a late period. Four men lived in an "A" tent, which was fastened to layers of logs about three feet in height by means of a chimney built of sticks and adhesive mud, and surmounted by a barrel.

No class of persons follow a life of greater indolence than soldiers who are living in winter quarters;[27] and the camp afforded good opportunities to study human nature, and learn the motives which actuate many that enlist in the service. The regiment formed a hollow square in the evening, after dress parade; and the chaplains, in compliance with orders, offered prayers to promote the spiritual welfare of the troops. Those officers who seldom held any religious services upon the Sabbath, and employed their time in cooking, and swindling the men and poor negroes, were not listened to with any feelings of respect. The author was present upon these occasions in a certain command, and noticed that the appearance of the chaplain, who had acquired the sobriquet of "good and holy man," on account of the peculiar manner in which he performed his duties, was greeted with oaths by the line; and the epithets, "hypocrite," "pies," "rascal," "turkeys," and similar words, fell upon the ear. One colonel sometimes reproved his chaplain by saying to him, in strong language, "If you don't cook a better dinner[28] than this

725]

27 The boring camp life was getting on the nerves of some of the more excitable officers. Regimental Adjutant, Lt. Walter Smith, turned in Capt. William Allen for being absent from an early morning roll call. Asked to explain, Allen wrote that he and Roberts were both present "and if that traveling whiskey cask Lt. Smith says we were not, he is a damned liar." Tried for "conduct unbecoming an officer," Allen was convicted, and sentenced to forfeit one month's pay. [NARA file 807]

28 Chaplain Watson "provided and cooked meals for the field and staff officers of the 11th, and Col. Blaisdell threatened several times to punish him when dinner was not satisfactory by ordering him to preach on the following Sunday." [Memoirs] The epithet "pies" seems to be a variation

to-morrow, I will have you tied to the flag-staff next Sunday, and make you preach two hours to the regiment." Threats like these made the offender redouble his efforts; and, in his anxiety to provide a good meal for his mess upon a certain day, he neglected to attend the funeral of a private; and the surgeon read the burial service at the grave.

Some of the citizens with treasonable proclivities, who resided near the camps, demanded large amounts for alleged damages to their property by the soldiers, when they had no just claim. A captain in the regiment purchased some boards in a shed for twenty dollars; but the owner declined to receive the money when it was tendered, and remarked that it was "all right," and he did not wish for any compensation. The officer subsequently ascertained that the scoundrel presented a bill, amounting to three hundred dollars, against the Government, for injuries which he had sustained by the loss of his lumber. Another party filed a long account for damages to a saw-mill; and, after his death, the son and administrator of the estate trebled the sum. When the premises were examined, it was clearly proved that the building was tumbling to pieces on account of its age, and no use had been made of it for years previous to the arrival of the division.

Negroes continually escaped, and were concealed and sheltered in the quarters; and disturbances always occurred whenever their masters, in compliance with instructions from army headquarters, appeared to search the camps for the purpose of seizing and carrying them to their old cabins. The slave-hunters were forced to leave some regiments by the indignant soldiers, who threatened to shoot them if they persisted in entering their lines. Besides this unfortunate race, refugees and

---

of "pious" in its negative connotation.

deserters occasionally passed across the Potomac from Virginia by means of logs, rafts, and "dug-outs."

The plain rations furnished by the Government did not tempt the palate; but the supply largely exceeded the demand: and foreign officers who visited the camps stated that another force, of the same number, could be well fed by issuing to them the food that was wasted. The enlisted men who had formerly served in European armies often made the same assertion. An English author [Anthony Trollope], who inherited from his literary mother a bitter enmity against the United States,[29] says, "The great boast of this army was that they ate meat twice a day, and that their daily supply of bread was more than they could consume." The wife of an officer, who noticed with much interest the quality of food which the soldiers devoured at their meals, remarked, in a tone of surprise, "The privates fare well: they live just as they do at home in jail." The third brigade, comprising regiments from New Jersey,[30] joined the division; and, like all the troops from this State, their gallant conduct, during the years that the command existed, rendered invaluable aid to the national cause.

Three companies of the regiment, A, F, and K, under the command of Lieut.-Col. Tileston (who fell in the second battle of Bull Run), were ordered to march through the counties of Lower Maryland,[31] in the latter part of December, to search for arms, and prevent the passage of recruits, stores, and mails, from Maryland to Virginia. The detachment was absent three weeks upon this service; and the company remained a third of the time

---

29 Frances Milton Trollope wrote of her dislike for Americans in *Domestic Manners of the Americans* (1832).

30 5th through 8th New Jersey, commanded by Col. Samuel H. Starr, encamped just to the north of the Hooker Brigade.

31 The counties of St. Mary's, Calvert, and Charles.

at Piney Point, which was seventy miles from the camp at Budd's Ferry. The weather and roads were unfavorable for public travel: the troops bivouacked upon different nights in quarters between which there was a vast contrast; and occupied, according to circumstances, court-houses and hotels, or barns and straw-stacks. A detail demolished at Allen's Fresh a boat which had frequently crossed the Potomac, although the owner solemnly declared that it was "as innocent as a new-born lamb." A squadron of cavalry patrolled the bank of the river, and watched the numerous inlets and creeks; and the people treated the officers with extreme courtesy, and concealed, under a mask of hospitality, their inimical schemes. A major, who commanded the [cavalry] detachment, was especially welcomed by the family of one of the most wealthy and aristocratic citizens, and contracted an engagement to be married to his daughter. Yielding to the alluring entreaties of his betrothed, he overcame, by means of his position and representations, the objections of the loyal inhabitants, and procured the discharge of her rebel brother and a cousin from Fort Lafayette, in which they had been justly confined by the Government. As soon as these relatives were released from the walls of their prison, the artful lady, who, like all traitors, was destitute of every sentiment of honor, released the officer from his engagement, and treated him with the utmost contempt.[32]

The soldiers passed through the most thickly-settled slave-holding section of the State, in which, as a natural result, the mass of the people was ignorant, and attached to the cause of the enemy. The country had been settled at an early period; and some of Lord Baltimore's landmarks, which indicated the boundaries

---

32 The victim of this southern "treachery" was Maj. George H. Chapman, 3rd Indiana Cavalry, who monitored secessionist activity in lower Maryland for Hooker.

of his grants, were still standing. Like the M'Sweyn Family, described by Dr. Johnson,[33] there was no progress from one generation to the next. "For the son is exactly formed upon the father: what the father says, the son says; what the father looks, the son looks." The stagnation that existed in this locality was adapted for the man who wished to "immerse himself and his posterity for ages in barbarism." Many times, when the soldiers were passing a dwelling, the song of "My Maryland"[34] was played upon the piano and sung by the young members of the household; and wishes were uttered like the following; "I hope you will always be whipped, and have to run away as you did at Bull Run." In the church at Leonardtown, [candy] lozenges, upon which treasonable mottoes were printed, were scattered in the pews that they occupied. One citizen owned two savage mastiffs, which he had designated by the names of "Jeff Davis" and "Beauregard," as a tribute of respect for those rebels.[35]

The negroes were delighted to see the troops, and danced with joy when the officers placed in arrest, or examined the premises of "massa," who had hitherto swayed the community without opposition. They always told falsehoods, and made answers that they thought would shield them from brutal punishment if they were questioned in the presence of any white

---

33 Blake refers to Samuel Johnson's *A Journey to the Western Islands of Scotland* (1775) in which the McSweyn's are described as an example of highland hospitality.

34 Poem written by Maryland native James Ryder Randall in 1861 expressing the author's Confederate sympathies; set to the traditional tune "O Tannenbaum."

35 Pvt. Edward Gazet had a friendlier reception: "The people treated us first rate gave us dinner whenever we dropped in on them at mealtimes. I don't know if it was through fear or love for us and don't care much, but this I know that we got a good name and the man that least us the house gave the boys a parting drink and told us that if any troops come there again he hoped it would be Co. A of the 11th." [Gazet letters]

citizens; but gave all the information they possessed when assured of protection [by Union officers], and eagerly pointed out the places in which arms and colors had been concealed; and, by this means, an old cannon was discovered in a swamp. They were celebrating the [Christmas] holidays; and I observed with surprise, what my experience confirmed, that the majority of the whites and blacks of both sexes talked alike; and the peculiar dialect which marks the performance of Ethiopian minstrels is the common language of some of the best classes. It was generally impossible to distinguish between the master and his slave in the night, if the tones and style of the conversation formed the sole basis of judgment. "Who dat trow de snowball?" was the question asked by a very intelligent citizen of Budd's Ferry, who owned more than a hundred "hands," when a mischievous soldier, by skillfully throwing this missile, struck his portly back. The mulattoes, who had lost many of the physical features of the race, were numerous; and I saw near Leonardtown two men with sable complexions and bright auburn "wool." One of them, without knowing the fact, adopted an old saying,[36] in letter and spirit, and stated that he did not drink any whiskey or get drunk until he became a Roman, and did the same as the Romans. The often-repeated remark, "I's a Roman," was the expression which he used to show that he belonged to the Catholic Church. When the men referred to the peculiar color of his hair, he said that it "growed dar;" and he liked it because he had seen pictures of God, and "God had red har." This strange dogma of faith comprised the chief article of his religion, and was maintained with great fervor by allusions to the sacred paintings, the "pictures of God," upon the walls of his house of worship.

---

36 "When in Rome, do as the Romans do."

Large flocks of turkey-buzzards were constantly flying in the air; and a farmer informed me that they disappeared and flew to the field of carnage after the battle of Bull Run; from which they returned to their old haunts, near the bluffs of the Potomac, after an absence of three months. The excitement and exercise attending the march in the midst of winter improved the health of the men; and, during the three weeks that followed the date upon which they rejoined the regiment, not a single case of sickness occurred in this command, while the remaining companies had from thirty to fifty names upon the surgeon's list.

The natural obstacles of the climate, and winter season, prevented the entire army from advancing; and preparations for an active campaign were made when the spring approached.[37] Shelter tents, that beneficial invention which enables the soldier to transport his house upon his shoulders, and cartridges containing three buck-shot and a bullet,[38] were issued to the regiment; and surplus stores and clothing, which could not be carried, were sent to Washington and Alexandria for storage. Upon March 9, 1862, the enemy burned the schooners and steamer "George Page;" evacuated the earthworks upon which they had daily labored for six months; and the American colors were fastened to the rebel flag-staffs in the afternoon.[39] The reports of explosions in the abandoned forts and camps that were located in the interior gradually grew fainter; and the sounds were barely perceptible when the retreating forces were ten miles from the Potomac.

---

37 Despite approval of Hooker's plan to attack the Virginia batteries, McClellan never intended to implement it. A contemplated Union landing near Urbanna would outflank the Confederates and force the batteries to be abandoned.

38 Smoothbore musket "buck and ball" ammunition.

39 Johnston pulled the Confederate army back behind the Rappahannock River, effectively killing McClellan's Urbanna plan.

The river was once more free: mariners no longer preferred darkness to light, or anxiously scanned the forests and hills of Virginia to discover amidst the beauty of the scenery the "horrid flash" of the rifles and cannon of a lurking foe; and transports in large numbers hourly sailed by the silent forts. Most of the siege-guns in the batteries were over-charged, and burst [when fired]; and the carriages were burning when the Union took possession. By a fortuitous coincidence, one loaded piece, heated by the fire of the wood-work, discharged a ball at the moment that a schooner was within range; and this was the last solid shot which was thrown by rebel ordnance across the Lower Potomac. This unlooked-for explosion affected the tongues of the few soldiers who were naturally despondent, and they began to croak, "The rebels have not gone yet;" or "They are still there, and playing a deep game." When the appearance of the short-lived "Merrimac" alarmed the country, the barges and boats, which had been collected at Rum Point for the purpose of transporting the division to Virginia to assault the batteries,[40] were loaded with gravel and stones; and men were detailed to sink them in the channel if the iron monster attempted to ascend the river and bombard Washington.

Upon April 5, 1862, the troops, animated by the victories which had recently followed the Union armies,[41] cheerfully quit the familiar scenes of Budd's Ferry, bivouacked upon the banks, and embarked on the sixth upon the steamboat *Emperor* which

---

40 Lincoln was so exasperated by the blockade that on March 8 he issued a direct order to attack and eliminate the Potomac batteries. Confederate evacuation of northern Virginia started around the same time.

41 The capture of Forts Henry and Donelson, the Battle of Pea Ridge, and the capture of Island No. 10 in the Mississippi River. Also, Kentucky and most of Tennessee fell under Federal control.

conveyed the regiment during the succeeding week down the Potomac, and up the York.[42]

____

42 In October 1861 Col. William Blaisdell assumed command of the 11th
   upon the resignation of Col. Clark. Hooker was promoted to division
   command in early 1862, succeeded by Brig. Gen. Henry M. Naglee, a West
   Pointer who wasn't well received: "Our new brigade commander is
   inclined to carry things with a high hand. The other night he visited all the
   pickets and arrested thirteen officers whom he found asleep." [Wainwright,
   18]

Chapter III

The Siege of Yorktown

A SEVERE STORM prevailed during the voyage; and the vessel was compelled to anchor at Piney Point and Fortress Monroe, and occupied six days in sailing a distance, which, under ordinary circumstances, could have been easily made in twenty-four hours. The "Emperor" towed two schooners that carried artillery-horses, a number of which died from the effects of exposure to the weather. The boat was crowded to excess with soldiers and horses; the rations were insufficient for the unexpected length of the time that was consumed; the distressing pangs of sea-sickness affected a number; the first wave which washed the deck temporarily submerged one hundred and fifty men, who were sleeping in their blankets; and general discontent was produced by these combined causes.[1] A happy disembarkation took place upon April 12, at Ship Point, which was ten miles from Yorktown; and the regiment formed at this time a part of the first brigade, second division, third corps,[2] which were respectively commanded by Generals Naglee, Hooker, and Heintzelman. Strict orders had been issued that no negroes should be taken upon the transports at Budd's Ferry; but many who had escaped from servitude, and labored for officers, mysteriously rejoined the troops at this place.

---

1 Blake glosses over the magnitude of the Peninsula invasion: "121,500 men, 44 artillery batteries, 1,150 wagons, over 15,000 horses, and tons of equipment and supplies." [Sears 1999, 167]

2 The other two divisions of III Corps, Porter's and Hamilton's, had arrived at Fort Monroe on April 3. Hamilton was replaced by Brig. Gen. Philip Kearny on April 30. Porter's division transferred to V Corps in mid May.

The division marched through some abandoned rifle-pits and redoubts, and encamped in front of the works at Yorktown, within range of rebel cannon. The men were so compactly massed, that brigades occupied less space than a regiment in its regular camp. The commands of Generals Hooker and Kearney pitched their tents upon the ground that had been the headquarters of Washington and Lafayette in the Revolution; and the fires of patriotism glowed with greater intensity when the soldiers beheld the mounds which their ancestors had built to win success and national independence.[3] While they were upon the march from Ship Point, the provost-guard passed by the column with a squad of prisoners; and I saw among them a spy, who had visited the camps at Budd's Ferry, and distributed pious tracts. This entire country, from the landing-place to Yorktown, was a vast swamp: the roads, which were unfitted for general use, had been converted into canals by the frequent rains; and it was necessary to corduroy them before the wagon-trains and heavy artillery could proceed to the front. There were many stagnant pools upon the surface; the soil was filled with springs; and the companies obtained good water from wells which consisted of empty beef and pork barrels, which were sunk into holes two and three feet in depth.

The soldiers of Gen. Heintzelman's corps were engaged for the ensuing three weeks in fatiguing and incessant labor upon the works that were constructed under the direction of the engineers,[4] upon the right of the army; but some divisions that formed the reserve performed no service. The men were ordered to fall in

---

3   The surrender of Lord Cornwallis at Yorktown in October 1781 led to peace negotiations that resulted in the Treaty of Paris, 1783.

4   McClellan was the Union army's acknowledged expert on siege warfare, having witnessed the siege of Sevastopol in 1855. References and comparisons to Sevastopol appear often in the reports of the Yorktown operations. [OR XI/1, 348 and 356-358]

quietly in the morning, between three and four o'clock, to form a line of battle; and the ranks shivered in the cold mist that usually covered the earth, and remained in the rear of their stacks until sunrise. A majority of the regiment, and sometimes the whole of it, were daily detailed for fatigue-duty; and reported with arms and equipments at corps-headquarters, where every man was furnished with an axe or spade, and large working-parties proceeded to the different parts of the line.[5] It was officially stated that these detachments, furnished by the corps, comprised fifteen thousand soldiers, which was more that half of its effective strength; and the picket in its front required the unceasing vigilance of five thousand men. The first parallel was established near Wormley Creek, a sluggish stream of an irregular width that flowed between banks which were covered with a thick growth of timber; and the siege was prosecuted with untiring industry. Bridges and roads were constructed in the ravine for the passage of cannon and ammunition; and ditches, revetments, and parapets were built in the advance. The tract of country between the camps and the breastworks of the enemy was extremely level; but forests of pine and hemlock, and the absence of commanding hills, prevented the rebels from discerning the movements of the besieging forces. The balloon, which this state of facts rendered a necessity, made daily reconnoissances; and was stationed in a cavern, which seemed to have been prepared by Nature for this purpose, when it was not floating in the air. An intense excitement existed in the army

---

5 During the siege a special title was created – General of the Trenches – which rotated daily. Brig. Gen. Daniel Butterfield complained that "Five hundred men, under Lt. Col. Tileston of the 11th Massachusetts Volunteers, did not get to work until two hours after their proper time, owing, I think to lack of thorough understanding of the hour which they were to report and the exact location of the work they were to be engaged." [OR XI/1, 384-389]

upon the 28th [of April], when the ropes that were attached to the [gas] car parted, and the balloon, containing Gen. Fitz John Porter, rose rapidly, and was wafted towards Gloucester Point, until another current bore it in the opposite direction; and unfortunately for the country, and the reputation of this officer, it fell inside of the Union lines.[6] A division of labor was required to complete the works, and a small number felled and split trees; but the main body of the troops was employed in the trenches. An uninjured saw-mill, that was located a short distance from Gen. Heintzelman's headquarters, was continually in operation in the charge of soldiers, and furnished dimension-lumber to the engineers. After the first parallel had been finished in the solitude of the forest, regular approaches were made in the night, when they were invisible, and strengthened during the day, until the advanced works were erected upon the plain in front of Yorktown, within three hundred yards of the enemy, although batteries and mortars were planted at points which varied in their distance from the town from one-half to one and a half miles.

Aided by the darkness, a small force was cautiously deployed in a certain direction, and silently labored with their shovels, placing the excavated earth upon the side nearest the enemy; and daybreak revealed an extended trench, that was two or three feet in depth and width. The laborers, knowing that each shovelful of gravel increased the size of their shield and made their position more secure, worked with great diligence. The officers in charge of the fatigue-parties, surrounded by circumstances that excited or confused the mind, sometimes

---

6 Porter's official report does not refer to this episode but does indicate that he was favorably disposed toward balloon observation "if the observer be intelligent and educated in the military profession." [OR XI/1, 312] Correspondence between Prof. Lowe and McClellan indicates that the incident may have taken place earlier (April 11) than indicated by Blake. [Sears 1988, 54]

committed queer mistakes. Upon one occasion, the gravel was placed upon the wrong side of the ditch, and the regiment, after it relieved the night detail, transferred the newly made bank to its proper position. The work did not connect upon the right in another portion of the line; and an extent of ground ten yards in breadth was exposed to the rifles of two rebel sharpshooters, who had climbed up in the large chimney of a burnt house, and made a loophole by knocking out one of the bricks. The soldiers, keeping close to the earth, rushed over the dangerous spot in one rank to the scene of their toil. These narrow trenches were enlarged until they were ten or twelve feet in width, miles in length, and four or six feet deep; and other bodies of troops built traverses and magazines, and transported the artillery and necessary supplies to the batteries that were completed. The technical terms used by engineers in giving instructions were remembered by the men, who always listened with confidence and respect to the comments of this superior class of officers, and in conversation facetiously and glibly — *Talked of rampart and ravine, And trenches fenced with gabion and fascine.*[7]

The hostile gunners and pickets saw the earth when it was thrown upon the parapet by thousands of revolving spades, and attempted to obstruct the progress of the siege; but their constant efforts and volleys of shells and Minie balls tested the works, and the small loss that was sustained during their construction showed that they were properly executed. The soil, in many places, was composed of minute marine shells; and the soldiers exhumed new specimens in widening the trenches, and faithfully served their country, and, at the same time, gratified their taste for geological studies. The redoubts and rifle-pits of the Revolution, which had diminished until they were only twenty

---

7 *The Pilgrim of Glencoe* by Thomas Campbell (1842).

inches in height, intersected those of the Union army at several points. A few metallic relics, corroded by the rust of eighty years, were brought forth from their hiding placed in the earth. The workmen became accustomed to the concussion of artillery, and the harmless results that followed a large proportion of the reports dispelled every emotion of fear; and many amusing incidents illustrating the good spirits of the troops occurred in the fortifications. Questions relating to the claim of two soldiers to use the same spade often arose; and, after the usual amount of reason and profanity had been exhausted, blows were exchanged, and several personal encounters took place, while the missiles of destruction were flying over the combatants. If a hat and blue overcoat were placed upon the handle of a spade, and elevated above the bank, bullets always greeted them; and sometimes, when the clothing was removed, the rebels who perceived the motives of the invisible soldier that held it fired to enable the men to witness their skill as sharpshooters. One of the Union marksmen saw by means of his telescopic rifle a man upon the ramparts of Yorktown, who amused his companions by making significant gestures towards the lines, and performed queer flourishes with his fingers, thumbs, and nose. The distance between them was so great, that the buffoon supposed he was safe; but the unerring ball pierced his heart, and he fell inside of the works. The brigade commander [Naglee] called for volunteers to dig a sharpshooter's pit in a dangerous position in the front, and excused two men in the regiment who performed the task from fatigue duty for the space of three days.

At the end of two weeks, an important advance had been made from the first parallel; and the massive breast-works of the enemy, upon which guns of different calibers had been mounted, could be examined. When their artillery had a good range upon an unfinished work, a man was constantly upon the watch, and shouted, "Lie low!" or "Look out for that shell!" as soon as the puff of white smoke darted forth, and preceded by a few precious seconds the arrival of the iron messenger. Pick-axes and shovels were instantly cast upon the ground; while those that used them jumped into safe ditches, and promptly resumed their labor when the shot passed over them, or the sentinel exclaimed, "They fired that ball the other way." An officer who was engaged upon this duty said, "There is a big cloud, and that hundred-pound gun has burst," when one of their most powerful pieces had been shattered into useless fragments. There was more danger in returning to the camp after the allotted task for the day was ended, then in toiling at the front; because the foe, who had ascertained the hour at which the "shovel divisions" were generally relieved, opened their batteries, and scattered shot and shell into every portion of the road over which they were obliged to march.

The proper discharge of picket-duty at this time required soldiers who possessed, in the highest degree, the qualities of courage and self-possession. The principal part of the regiment rested during the day in a ravine which partially sheltered it from the fire of the enemy; and remained in a state of readiness to rush at a second's notice, to the point that was attacked, whenever those upon posts crouched behind trees and knolls gave the alarm. The sharpshooters of both armies, who were concealed in little pits in the extreme front, were always relieved in the night; and the person who moved upon the open plain was a target for deadly rifles. There was no relief for those that were wounded in these hazardous positions, until the sun sank beneath the horizon; and some of these unfortunate heroes languished and died, while their comrades were unable to alleviate their sufferings. It was a difficult task for the officers to restrain the curiosity of restless men who wished to reconnoitre the rebel works, — which resembled in their appearance the bank of a railroad, — count the cannons, and recklessly expose themselves for this purpose; but the command, "Get down there, you fool!" or, "Stay by the stack in your places!" made them return to the ranks. The chariot of Phœbus[8] traveled slowly through the skies; and during these weary hours, letters were written, books were read: but the largest number, exhausted by unremitting labor upon the forts, slept near their muskets, and were undisturbed by the flight of shells and bullets.

The companies advanced about two hundred yards from the reserve after sunset, and deployed upon the field in front of the ravine, and groups of three men were stationed at frequent intervals. Night had blinded the eyes of the attentive sharp-shooters; but their ears caught every sound, and the slight crackling of dry twigs and bushes beneath the cautious feet caused two or three balls to whistle in the vicinity. Enveloped in a rubber blanket, and resting upon their breasts on the earth, two men upon each post stimulated to the utmost their powers of observation, while the third — who was regularly relieved by a comrade — slept as well as the weather and circumstances would permit. If a light was seen, it kindled a fire of musketry which extinguished it; and although nine-tenths of the soldiers, like ordinary people, were in the habit of stating that they could not live without their pipe, tobacco, and a "smoke," the rifles of the rebels effected a temporary reform in this respect, and there was no smoking upon picket-duty in the front. It seemed superfluous, but strict orders were issued prohibiting fires: and sometimes a reckless man, protected by the trunk of a tree, tied a match to a stick, and held it up to attract the notice of the enemy; and the experiment was always successful. Myriads of the insect termed,

---

8 The chariot of the sun is driven by Phœbus (Apollo), Greek god of light and poetry. [Hamilton]

in common language, the firefly, generally filled the air; and a field-officer, observing one of them upon a bush in a very dark night, and supposing that is was the ignited match of an inveterate smoker, said, in a low tone of voice, "Put out that light!" The order was unobeyed, and he again spoke: "Put out that light, you scoundrel!" and then moved towards the object that had excited his indignation, and frightened the fly, which flew away amidst the half-repressed laughter of the spectators.

The corporal of a post that I relieved the first time the regiment was upon picket at Yorktown, pointing to a hole which had been made by the bursting of a shell about twenty feet from him, jocosely said, "There is a grave already dug, large enough for three of you." The national forces were not allowed to build camp-fires in the night, because it was thought that the rebels would gain, by this means, information regarding their number and position; but the enemy exhibited no such fear, and the heavens above Yorktown and Gloucester Point reflected the lights that extended four or five miles in the rear of their works. While the rumbling of the artillery, and wagons conveying supplies in the distant encampments of both armies, the sharp stroke of the axe and the dull ring of the spade upon the entrenchments, and especially the shrill clatter of the machinery of the saw-mill, confused the ear with opposing sounds, and unnatural stillness prevailed in the space of ground between the hostile pickets, who were so near each other at times, that the rattling of an officer's sword, and the click of a rifle when it was primed, were distinctly heard. As the Union officers expected a sortie, and the commander of the foe anticipated an assault, the tour of night duty rarely passed away unmarked by heavy firing; and volleys of musketry, and charges of grape, swept over the field, and a decisive battle seemed to be imminent. The midnight air chilled the soldiers, who were obliged to remain quietly upon their posts, trembling with cold; and there was so much suffering in the few hours of picket-service, that they were willing to become disciples of Zoroaster, and worship Ormuzd[9] when the twilight gladdened their hearts, and they retired to the ravine.

The bushes and short pines affected by the breeze, and flickering lights of the distant camp-fires, appeared to some excited eyes like advancing horses, and masses of men; and shots were fired at the waving branches. Negroes and small squads of deserters, and soldiers who lost their way in attempting to pass between the posts, came inside the lines every night; and orders were given to the sentinels to allow them to approach, but to resist a large force. One rebel, who was taken by the regiment, gave an amusing account of the means by which he effected his escape. A man challenged, when he heard the sound of footsteps, and, receiving no answer, fired, and afterwards boasted that he had killed a "Yank;" but his comrades saw a dead pig in front of his post upon the next morning. The pickets, deterred by the boundless ridicule which this incident caused in their brigade, were extremely cautious, and did not discharge their pieces in the night, unless they were absolutely certain that the enemy menaced the line; and this prisoner crept upon his hands and knees until he had "grunted by the guards."

The troops improved their shelter tents, the ridge-pole of which was scarcely three feet from the earth, by making doors and carpets of boughs and strips of bark, which were taken from the trees by the use of the bayonet. At the head of the company street were the headquarters of the brigade commander, — a general of great wealth,[10] who carried some pullets upon the

---

9 Zoroaster (7th century BCE) was a Persian religious teacher and prophet who taught the existence of a supreme deity, Ormuzd (Ahura Mazda), the source of all light and purity, and emblematic of the sun.

10 Gen. Naglee had been a California banker before the war. He continued to

campaign; and the crowing of the rooster in the morning gave a domestic character to the camp. Although the batteries of Magruder hourly opened, our gunners, in those [gun galleries] that were completed, with the exception of that upon the extreme right,[11] were commanded not to return the fire, unless attacked; and the [rebel] garrison was slightly molested during the siege. The bands were silent, and no bugle or drum sounded the calls, which might disclose to an observing foe the dispositions of the Union army. The Prince de Joinville, who says in his pamphlet[12] that he "used to go to the front for this cannonade, as if it were an entertainment," makes the following strange statement, which is poetical language and fiction: "On fine spring evenings, the troops came in gayly to the sound of martial music, through the blossoming woods." The soldiers, who did not "go" from a place of safety in the rear, but lived at the front, who were present, not for "entertainment," but to perform the hard labor and sleepless vigil, will say that the prince's description is as accurate as his painting of the battle of Gaines Mill.[13]

The dead and wounded were generally carried to the rear, upon the Yorktown Road, which ran by the camp; and the

---

alienate his fellow officers on the Peninsula. "Naglee feels his shoulder straps too much, and forgets that they do not give him command outside his own brigade." [Wainwright, 39] Later in the month, Hooker got Naglee transferred and replaced by another West Pointer: Brig. Gen. Cuvier Grover.

11 The battery on the extreme right was Battery No. 1 – Brig. Gen. William F. Barry wrote that it was the only one that "circumstances permitted to be fired." He lamented that the Confederate retreat from Yorktown "deprived the artillery of the Army of the Potomac of the opportunity of exhibiting the superior power and efficiency of the unusually heavy metal used in the siege." [OR XI/1, 348]

12 French nobleman on McClellan's staff. The "pamphlet" is *The Army of the Potomac, Its Organization, Its Commander, and Its Campaigns*.

13 Joinville's painting is more a group portrait of McClellan's staff than a depiction of the battle. Original text says "Gaines Hill."

surgeons of the brigade amputated the arm of one of the engineers, who had been injured by the fragment of a shell while he was reconnoitring the forts. The operation was highly successful; and the officer remarked, in the best of spirits, "I finished my sketch." He remained at brigade headquarters about two days, and then entered a hospital which was under the supervision of regular army surgeons. They expressed the usual contemptuous feelings for volunteers; called the work a "botch;" removed the old bandages; and needlessly experimented upon the limb, until secondary hemorrhage took place, and within six hours the person was dead. Great indignation was expressed by the members of the brigade, when they learned the facts attending his death; and those who were well acquainted with the case asserted that he had been murdered.

The constant interruption of sleep by the artillery and musketry; the formation of the line of battle at all hours of the day and night; the continued labor upon the earthworks and roads; the exposure and excitement in the camp and upon picket-duty; the rain which fell two days in three, and increased tenfold the burdens of the troops; the quality of army rations, and the absence of medical supplies; the lack of time and means to preserve the habits of cleanliness; the swampy nature of the country, and the character of the climate, — produced disease, and thousands were afflicted with fevers. The list of patients in the hospitals included a number of cowards, who always disgrace every regiment, and feigned sickness to escape the dangers of the siege. The contents of two quart bottles and two pint boxes comprised the medical stores of many commands, and were administered as a specific for all complaints. I have seen a surgeon give medicine from the same cup for a sore throat and a scalded foot.

An hour before daybreak, upon April 26, two companies from the regiment, and a detachment from the First Massachusetts Volunteers, captured and leveled a lunette[14] which was built in an advanced position and annoyed the working parties. I was awakened about two hours after midnight by the stern voice of the captain of an adjoining company, which he was forming into line with difficulty, because some of the men wished to drink the coffee which had been made for this occasion. He summarily ended this cause of delay by kicking over their cups; and they marched with sullen steps to the front, but performed their duty, and received the thanks of the commander-in-chief.[15]

The preparations for the bombardment were nearly completed; and it was stated that one hundred and one cannons and mortars, some of which were two-hundred pound Parrott guns, mounted in fourteen batteries, would open upon the enemy on May 5, and throw into the rebel works a shell in every second.[16] The [rebel] cannonading was unusually active during the afternoon and evening of [May 3rd]; shells were constantly

---

14 The companies involved were A, H, and I (1st Massachusetts), plus A and G (11th Massachusetts). "The bayonet attack on the lunette was made by Company H, which suffered four killed and twelve wounded, taking fourteen prisoners, all from Company E, 19th Virginia. The 11th's Company G, carrying picks and shovels, was assigned the task of leveling the works." *Chelsea Telegraph and Pioneer*, May 3, 1862.

15 McClellan sent a special report to Secretary of War Stanton and received in return a letter calling it a "brilliant affair." [OR XI/1, 382-384] The northern press, eager to report on any military success by the Army of the Potomac, exaggerated the significance of the lunette attack.

16 Barry itemized the guns: two 200-pound Parrott guns, eleven 100-pound Parrott guns, thirteen 30-pound Parrott guns, twenty-two 20-pound Parrott guns, ten 4½-inch siege guns, ten 13-inch seacoast mortars, five 8-inch siege mortars, and three 8-inch siege howitzers. "Three field batteries of 12-pounders were likewise made use of as guns of position." [OR XI/1, 339]

exploded in the camps, with few serious results; and a desperate sortie was considered a sure event. The morning of [May] 4th was quiet; and the pickets discovered that the furious artillery fire had deceived the general, while Yorktown had been evacuated, and the national flag was triumphantly placed upon the abandoned forts. This was an event that had been wholly unforeseen. The fatigue-details labored in the trenches during the following morning. Loud cheers resounded along the line, from the York River to Warwick Creek, when the result was officially announced; and the bands, which had been dumb so long, again enlivened the soldiers; and the notes of a thousand drums, fifes, and bugles, filled the woods with a "discord of melody." The division, led by Gen. Hooker, and forming the infantry advance of the army, with rations for three days, and in heavy marching order, pushed forward at noon upon the Yorktown Road to support the cavalry which was pursuing the retreating forces. In passing by the tenantless pits of the sharpshooters upon the plain, the man who had firmly grasped his rifle, and crept silently from post to post during the perilous nights of the siege, viewed the harmless works, and the ruins of the storehouses and wharves which were burning at this time with emotions of joy; and —
Full well he bore his knapsack unoppressed, And marched with soldier-like erected crest.[17]

Thousands of slaves had labored for months upon these [rebel] fortifications, which had been designed by skillful engineers, and had "formidable profiles, eighteen feet thickness of parapet, and generally ten feet depth of ditch," "with well-made sod revetments."[18] The peculiar spelling of the inscriptions,

---

17 *The Pilgrim of Glencoe* (Thomas Campbell). Blake ends the quote with "breast" but the poem actually says "crest."

18 Blake is quoting Brig. Gen. John G. Barnard, Chief Engineer of the Army of the Potomac, whose report was later published in OR XI/1, 316.

and especially the word "dide," caused much amusement in the ranks when they passed by the graves of some rebel soldiers.

Before the evacuation, the rebels buried torpedoes in the vicinity of the springs, hospitals, and other places which they supposed the soldiers would visit, and for a distance of three miles in the roads over which the army marched. Some of those who first entered the works were killed and mangled by this diabolical means. The newly made earth revealed the location of these concealed infernal machines; and pioneers drove stakes into the ground, and guards were placed near them to caution the troops and prevent them from walking upon destruction. Canteens, articles of clothing, and equipments were thrown aside [by the retreating rebels] to tempt the unwary relic-seeker; but the person who picked them up pulled the wire or cord which was fastened to the cap of a hidden shell. "Keep to the right!" "Go to the left!" "Don't touch that coat!" the sentinels shouted to the column as it moved forward, and vacillated from one side of the way to the other to avoid the serious consequences that would follow a misstep.[19]

---

19 The mines and booby traps were the work of Gen. Gabriel J. Rains, who had first used them in the Seminole War. Rains' corps commander, Gen. James Longstreet, heard about the practice and declared it "not proper and effective method of war." The dispute went to the Secretary of War, who ruled generally in Rains' favor, provided that the mines were so designed to "achieve a definite military advantage" and not indiscriminate harassment. [Perry, 20-27]

Chapter IV

The Battle of Williamsburg

THE TROOPS continued to march upon the Williamsburg Road, after leaving Yorktown, until they reached the "Half-way House," which was seven miles from the two places. They were compelled to halt three hours, because Gen. [William] Smith's division, that had moved upon another and more direct road, occupied the highway which Gen. Hooker wished to use. A squad of a dozen hatless and horseless cavalry brought the news that the enemy was in line of battle at a point about five miles distant; and this was joyful intelligence to regiments that were impatient to be baptized with the fire of conflict. An aide shouted in a loud voice, at sunset, "Gen. Smith's division will take the road to right, and Gen. Hooker will move to the left!" The black clouds that overcast the sky, and the gloom that pervaded the forest, *Made that darker Which was dark enough before* and the outcries of many *A soldier full of strange oaths[1]* were the only guides which were safely followed by the men, who pursued their course over a narrow pathway that led from Cheesecake Church[2] through a swamp, and frequently stumbled over logs, or sank into the water and mud. In the confusion which these circumstances produced, the woods were filled with soldiers, who were trying to find the way to the front; and one corps commander, with a portion of his staff, was isolated from the troops, and forced to remain absent until daybreak. Overcome by the fatigues of the march, the members of the division threw themselves upon the

---

1 *As You Like It*, Act 2, Scene vii (Shakespeare)

2 Hampton Parish Church. Cheesecake is a corruption of the native name Chiskiack. The church was torn down during the Civil War; its bricks used by Union officers for camp chimneys. [Hastings, 129]

ground a short time before midnight, and rested near the places in which they stood when the halt was ordered, until a heavy rain fell, which destroyed sleep, and increased the weight of knapsacks and clothing.

The lines were quietly formed upon May 5,[3] at daybreak; and the brigade [now under command of Brig. Gen. Cuvier Grover] advanced, in the midst of the storm, which still continued, upon the Hampton or Lee's Mill Road, that had been cut through a dense forest. A small earth-work, which the enemy held against the attack of the cavalry upon the previous day, had been abandoned in the night; and a large proportion of the command did not imagine that there would be any fighting until the reports of the skirmishers' rifles were heard in the front, and surgeons, chaplains, commissaries, quartermasters, cooks, and the horses of officers who dismounted, promptly retired to the rear. The brigade halted, while preparations were made for an engagement; and Gen. Hooker, who assigned the regiments to their posts, ordered the "Eleventh" to file to the right of the road, and smilingly said to the major [Porter Tripp], "This is a strong position, and the devil himself cannot drive you out of it." — "We are willing that he should try it," the officer replied. The firing between the skirmishers became animated: but the cartridges in the muskets had been wet by the storm, although every exertion had been made to prevent this result; and the snapping of percussion-caps was more frequent than the whistling of bullets. Whenever there was a cessation in the rain during the remainder of the day, the men were continually drawing their ruined charges, and cleaning the guns in the turmoil of the battle.

------

3 Original text says "May 8."

The topographical features of the scene of conflict, which was confined to a small extent of ground that bordered upon the Hampton Road, were similar to the those of Yorktown. Redoubts which extended across the Peninsula had been built upon a plain; and the open space, half a mile in width, in front of them, in which many pits had been dug for the sharpshooters, was bounded by the "forest primeval,"[4] a large belt of which had been felled, and formed an abattis that no organized body of troops could penetrate. The only means of approaching these works consisted of two roads, the Hampton and the Yorktown, which were exposed to the guns of the largest redoubt, called Fort Magruder. The regiments were deployed upon the right and left of the Hampton Road: the skirmishers steadily advanced,[5] while the enemy fell back until the men held a line in the edge of the abattis; and the sharpshooters picked off the gunners in the work. The thick mist which covered the earth in the morning confounded the rebels, who could not see the [Federal] battalions, and did not open their batteries because they did not know the point to which they could direct their fire with effect, and occasionally threw a random shot to elicit a reply from the artillery. A regular battery was unlimbered upon the plain after a

---

4   *Evangeline* (1847) by Henry Wadsworth Longfellow.

5   Blaisdell reported: "I arrived with the regiment in front of the enemy at fifteen minutes before five in the morning, and was ordered by General Grover to advance and deploy to the right of the Second New Hampshire regiment as skirmishers. As soon as I became unmasked my right company engaged with the enemy's skirmishers and reserve. A couple of well-directed volleys from Company E, Captain Bigelow, sent the enemy back in double-quick. On moving farther to the right, unmasking the whole regiment, I found a large force of the enemy's skirmishers, and immediately ordered Companies E, Captain Bigelow, and I, Lieutenant Robertson commanding, to deploy as skirmishers and engage the enemy at once, which was promptly and gallantly executed, the men advancing to within 300 yards of the enemy's works, driving all before them, and holding the position until 9.30 o'clock." [OR XI/1, 465-476]

delay caused by the bad condition of the roads; but the foe, ascertaining its position by the rattling of the wheels, immediately concentrated their fire upon it with such deadly accuracy, that the frightened cannoneers, with a few conspicuous exceptions, deserted their pieces[6] before a single shell had been discharged. The officers and men belonging to a battery composed of volunteers instantly rushed to the guns,[7] and, with the aid of another body of artillery, silenced every cannon in Fort Magruder after firing an hour. When the rain stopped at one time, the skirmishers reported that two white flags were visible upon the rebel parapet; but they were the colors of war, not peace.[8]

The regiment was ordered to skirmish[9] to the Yorktown Road at nine o'clock to ascertain the position of the enemy in its neighborhood; and no force was encountered, so that the connection between the front and the main portion of the Army of the Potomac was unobstructed. The second and third, the Sickles or Excelsior, and the Jersey brigades of the division were

---

6   Hooker noted that Webber's Battery H, 1st U.S. "was thrown forward in advance of the felled timber and brought into action in a cleared field on the right of the road and distant from Fort Magruder about 700 yards. No sooner had it emerged from the forest … than four guns from Fort Magruder opened on it, and after it was still farther up the road they received fire from two additional guns from the redoubt on the left. However, it was pushed on, and before it was brought into action two officers and two privates had been shot down, and before a single piece of the battery had been discharged its cannoneers had been driven from it despite the skill and activity of my sharpshooters in picking off the rebel gunners." [OR XI/1, 466]

7   The volunteers were from Battery D, 1st New York. Wainwright reported that "Every cannoneer at once sprang to the front, and headed by their officers, opened fire from four of Battery H's guns, while at the same time Captain Webber got some 15 or 18 of his men at the other two." [OR XI/1, 470]

8   Possibly "The Stainless Banner" – which was mostly white. It was not officially adopted as the national flag until 1863.

9   Grover had 1st Massachusetts skirmishing on his left while 11th

posted upon the left of the Hampton Road as soon as they arrived, and the firing at noon became a prolonged volley of musketry. The rebel commander, deceived by the length of the skirmish line and the vigor of the attack in the morning, supposed that the entire force of Gen. McClellan confronted him, and remained within the breastworks, and acted upon the defensive. Gen. Johnston wished to detain the Union columns at Williamsburg until night, to enable his trains to escape: all the retreating [rebel] divisions were halted when the skirmishers were driven in by the brigade; and some troops that were ten miles beyond the town countermarched,[10] and took an active part in the conflict during the afternoon. Unfortunately for the sacred cause, no such energy was exhibited by the commander of the Union forces, who styled himself the "senior officer upon the field,"[11] and declined to advance his brigades and make a feint, or re-enforce Gen. Hooker upon the Yorktown Road, although he was constantly notified in regard to the dangerous position which this officer was fighting to maintain.

The regiment remained in its position, near the Yorktown Road,[12] supporting a battery; and fixed bayonets when a charge

---

Massachusetts and 26th Pennsylvania did the same on his right. "In these positions with my whole force closely engaged and pressing forward, the fight was severe and unceasing, and our position without support until about 11 o'clock, when General Patterson's brigade made a timely arrival to meet a heavy force which was moving upon us through the timber on our left, evidently with a view to turn our flank and cut off communications with the rear." [OR XI/1, 473]

10 D. H. Hill's division of four brigades did the countermarch.

11 Brig. Gen. Edwin V. Sumner was the senior officer present but his corps (II) was not. McClellan had designated him commander *pro tem*.

12 Blaisdell reported that he "received an order from General Hooker to penetrate the woods to the right and rear to ascertain if there was any enemy between us and General Sumner, and if so, wipe them out." [OR XI/1, 476] He obeyed this order and returned to his previous position noting, as did Blake, that they fixed bayonets to repel a threatened attack

was expected from the cavalry which had assembled in force. Re-enforcements were hourly increasing the numbers of the enemy in front; and the extended lines of the division, which was pressed at all points, slowly contracted, while the foe moved a heavy body in the ravines which ran across the plain and sheltered it from the infantry fire, and attempted to turn the left by a vigorous assault upon that flank. To meet this mass that was advancing to annihilate the second and third brigades, the first, which was commanded by Gen. Grover, was withdrawn from the right, and ordered to support the left,[13] which was yielding gradually but making a stubborn defense. As soon as this movement was perceived, the rebels captured, without opposition, four guns, which were embedded in the mud, and could not be removed or guarded by the troops that were required for a more important duty in another part of the field. The regiment rejoined the division at the critical moment of the contest; and, while a new line was established, I witnessed one of the rare exhibitions of the power of a commanding presence, which great exigencies demand. The remnant of a brigade, which had resisted with brilliant valor the onset of superior numbers, discouraged by its large losses in officers and men and the absence of re-enforcements, retreated to escape capture; and the regiments mingled together in confusion while they fell back into the road. The yells of the exulting rebels proclaimed their success; and the gallant soldiers, who had taken the cartridges

---

on the artillery they were supporting.

13 Blaisdell then reported: "At about 3.30 o'clock I was ordered by General Hooker's assistant adjutant-general to march the regiment back to the left of the road and support the New Jersey brigade, which I obeyed as soon as possible, and arriving on the ground I became immediately engaged with the enemy, who was endeavoring to turn our left flank, continuing to hold them in until I was ordered to retire across the road in order that fresh troops might occupy our place." [Ibid.]

from the boxes of dead and wounded comrades when their ammunition was exhausted, commenced to rush to the rear in disorder. Gen. Hooker, who was riding along the lines, at once halted his favorite white horse in the midst of the medley, and exclaimed, "Men! what does this mean? You must hold your ground!" The voice that uttered these simple words had always taught justice and patriotism in the camps; the uplifted hand had always returned the salute of every soldier in his division; the form had ever been seen in the front when the storm of bullets fell and spared not; the dress was the uniform of a brigadier-general, who welcomed the dangers that belonged to his rank. The recollection of these exalted qualities flashed through the minds of all, and the commanding appearance was that of one who was "every inch"[14] a general. It inspired the timid with courage; the weak became strong; and every man stopped in his place, and faced the enemy.

Smith's [4th] New York battery of Napoleon guns was literally planted in the Hampton Road, which the rain and travel had converted into a bed of mud, in which the wheels and carriages were partially buried; and the cannoneers sank to their knees while they were loading and discharging their pieces. The abattis was a stumbling-block that impeded the advance of those who had felled it to check the national army; and the enemy was forced to make a long detour to the left to avoid it. The regiment crossed the road, and formed a line of battle which was parallel with it, and supported the battery that was double-charged with canister; and silently waited for the grand assault, which was every moment anticipated. One captain, who had always been excessively particular in dressing his company in the camp, and

---

14 Alluding to *King Lear*, Act IV, Scene v (Shakespeare).

was privately known as "Right Dress,"[15] displayed upon this trying occasion the most minute care; and henceforth there was no complaint about his conduct in this respect. The ranks in filing into the woods were transposed, so that the rear rank was in front when they halted; and some, who found themselves in this position in the rear, immediately forced their way into the rank that was nearest the foe. The hostile batteries, which had remained silent since nine, A.M., re-opened; and shot and shell swept the roads and woods; enfiladed the regimental line; and a number were seriously injured by falling limbs of trees that had been severed by them in their flight.

It was impossible to see objects with distinctness through the underbrush and huge oaks of the forest; but the ominous cessation of the rebel musketry and yells at this point indicated the movement of the troops, for which the general had already prepared with the limited force under his command. Subsequent events showed that Johnston intended to capture the battery by deploying two columns, which would subject the support to a fire upon its flank and front at the same instant. I perceived, through a slight opening in the woods, the ranks of the enemy, moving, within the distance of sixty feet, in a direction that was parallel with the position of the regiment; but they crouched upon the ground, with their faces towards me, and carried their rifles at "trail arms." Many of them were arrayed in the blue overcoats that had been taken from the dead and wounded when they succeeded in driving the brigades from the first line; and an earnest debate ensued, in which it was insisted that they belonged to regiments which had been fighting in the front; and officers of the highest rank ordered the men to hold their fire. "They are rebels!" "They are our own men!" "Don't you see those dirty

---

15 Probably Blake's own company commander, Capt. Benjamin Stone.

white hats?" "Those are our overcoats, any how!" "They are getting into line!" "They would not be so near if they were rebels!" were some of the outcries which were spoken upon every side. "I'll fire, orders or no orders!" said several soldiers; but, when they aimed, their companions, who supposed they were firing upon their friends, grasped them, and forcibly pointed the muzzles of the guns toward the skies. These moments of suspense, when hesitation was death, were agonizing in the extreme. Casual glimpses through the woods convinced the men of their real character; but there was no danger while the enemy was in motion in their front: and every person in the alignment placed his musket against his shoulder, and, taking a careful aim, anxiously awaited the final order of command, "Fire!"

"There is a white flag; don't shoot now!" "They are coming in to give themselves up!" were the exclamations of many tongues; when the color-bearer unfurled a small white battle-flag upon which a coat of arms had been formerly painted, which had faded so that there was a slight stain in the centre of it, which I noticed from my standpoint; but it would be invisible at the usual distance between contending armies. These colors were borne by the foe to deceive the soldiers; and the base purpose was successful in many conflicts, in which charges affecting the honor of this mode of warfare could be sustained by the testimony of thousands.[16] A private in the company upon my left walked about twenty feet towards the color-bearer to "show them the way to come in," and extended his hand to receive the white flag. A squad instantly discharged their rifles at him, and he fell

---

16 Pvt. Martin Hayes, 2nd New Hampshire, wrote: "Rebel prisoners with whom we afterwards conversed, alluded to the incident, but insisted that the flag … was only a battle flag faded until the figure upon it was hardly visible." [Haynes, 55]

upon the ground,[17] pierced by three balls; and the entire rebel line faced to the front. Every doubt vanished: and, before they could cock their pieces, the regiment was a wall of fire; and it was easier for the enemy to hew the way with swords through the abattis than overleap it. The proximity of the forces, the accuracy of the aim, and the perfection of the volley, produced in the rebel ranks a havoc which was seldom, if ever, surpassed in the history of war. "Shoot that officer on horseback!" a sergeant[18] shouted, pointing in the direction in which he aimed. A section fired: he fell from his horse; and, after the battle, there were fifteen or twenty soldiers who were certain that they had discharged the fatal shot. Many guns were foul and damp, and the cartridges were forced home by striking the rammers against stumps and trees during the act of loading. Not a bullet whistled from the front at the end of fifteen minutes: the attacking party fled, leaving more than half of their number upon the field; and the

---

17 Blaisdell commented upon the white flag incident: "a rebel officer displayed a white flag, crying out, "Don't fire on your friends," when I ordered "Cease firing," and Private Michael Doherty, of Company A, stepped forward to get the flag, and when near it the officer said to his men, 'Now give it them.' The men obeyed, firing and severely wounding Private Doherty, who immediately returned the fire, shooting the officer through the heart, thus rewarding him for his mean treachery." [OR XI/1, 476] Doherty survived his Williamsburg wound, only to be killed fourteen months later at Gettysburg.

18 The sergeant was Blake himself. [Memoirs]

large wounds made by the "buck and ball"[19] in this close action excited general astonishment.

The bullets from the enfilading column, in the mean while, decimated the regiment; and another change of its position was rendered necessary. The cannons of the battery, which the support had defended with such success, rewarded it by pouring into this advancing mass incessant charges of canister, which shattered the ranks to such an extent that they eagerly retreated to the ravine from which they had emerged. Jets of turbid water followed the recoil of the guns; and at night the generals and their commands were besmeared with the mud in which they were so often mired during the engagement. Gen. Heintzelman, the corps commander, joined Gen. Hooker at the front: but the first re-enforcements of infantry consisted of Gen. Berry's brigade, which was included in the division of Gen. Kearney, who had marched twelve miles; having left Yorktown in the rear of the army. While other officers bivouacked in the vicinity of the field, and rendered no assistance to those that had hitherto performed the fighting, the troops of Maine and Michigan, imbued with the spirit of their indomitable commanders, double-quicked through the slough of the Hampton Road; and no

---

19 The 11th was known as the "old mortar fleet" on account of their using a smooth-bore musket, with a cartridge of three buckshot and one musket ball. [Hutchinson, 32] Throughout the war, they used the Springfield Model 1816 musket (.69-caliber) – converted from flintlock to percussion-cap firing. Blake noted: "The Springfield rifle was superior to the smoothbore, but Colonel Blaisdell was opposed to any change and said the enemy would be within range before any battle ended." [Familiar Quotations]

Col. Robert McAllister, 11th New Jersey, wrote about the smoothbore musket when his regiment was first issued their arms: "We have got our arms at last. They are not what we wanted: the Springfield rifle musket. But we have got a new Austrian rifle for our flank companies and the Springfield smoothbore, buck-and-ball musket for the eight center companies. It is now thought that the musket with buck and ball is after all the best arm in the service." [McAllister, 209]

soldiers were ever more anxious to encounter inimical forces. "Holla! men, holla!" said Gen. Heintzelman, with his peculiar nasal twang, to the division which had been relieved, and was acting in the reserve: "Richmond is taken!" and the first Union cheers during the contest rose. He was not fully satisfied with the response, and shouted, "Bring up the bands! Play Yankee Doodle, or anything; but make some noise." A squad of musicians who belonged to different regiments was collected together, and the strains of Dixie and Yankee Doodle mingled in the din of the musketry and cannonade. The execution of the music in a public assembly, if it was viewed from an artistic point of view, would be pronounced inferior; but the effect upon the Federal battalions[20] was equal to the presence of a division or a battery. The soldiers, exhausted by the hardships of the preceding twenty four hours, received new strength; and the loud hurrahs misled a discomfited enemy, who did not make another advance after the repulse which Gen. Kearney's gallant troops had made decisive; and "the red field was won."[21]

The facts attending the death of two skulkers may be adduced to show the folly of trying to evade a soldier's duty in the day of battle. While one was peeping over a log, behind which he had concealed himself, a bullet entered his temple, which was the only part of his person that was exposed. Another, who had cautiously moved until he was ten yards in the rear of

---

20 Heintzelman's rallying cries were also noted by Pvt. Haynes of the 2nd New Hampshire: "To realize fully the ludicrous manner in which the sentences were snapped out, the reader should know that the old general was afflicted with a peculiar impediment in his speech, and that his words were run out as if on a wager to see how many he could get rid of in a certain time. Soon the bands struck up, the men who had no ammunition cheered like madmen, and the reckless spirit of General Heintzelman seemed to infuse itself through every body." [Haynes, 55-56]

21 *Marco Bozzaris* (1825) by Fitz-Greene Halleck.

his company, was pierced by two balls, which dodged between the legs of his comrades; and died after he had lingered in agony a fortnight. A captain in the regiment was compelled to quit the service because he used his eyes, and consulted a small pocket-compass, which led him, with a third of his company, towards Yorktown, when the reports of cannon, if he had pricked his ears, would have called him to the battle-field in the opposite direction.[22] However, upon the roll of honor, I saw the names of three brave men, reported killed, who would have been discharged by the surgeons for physical disability if they had not refused to accept the certificates to this effect. The bodies of some who were shot in the Hampton Road sank into the mud; and their remains were completely crushed beneath the wheels of the vast trains and heavy cannon that passed over them during the succeeding week.

The storm continued after the deadly struggle was ended; but it was not heeded by those who had battled from morning till night. In these hours of rest for the wearied, the non-combatants of the army, the members of the supply departments and ambulance corps, clerks, teamsters, musicians, and stragglers, who remain in the rear in time of action, visited the field, plundered the dead of both armies, and rifled the knapsacks of those who had [survived the fight] while they were asleep. The sun rose in a cloudless sky upon the 6th: but the enemy, having

---

22 Capt. John H. Davis was brought up on charges by Blaisdell for being absent during the Confederate attack at about 3:00 P.M. It appears he was separated from his company and went back up the road almost a mile to the hospital, as Surg. John W. Foye testified. Lt. Walter N. Smith and Lt. Charles C. Rivers testified to his absence, and Lt. Alonzo Coy, bringing up a train of ammunition, testified that when he passed Davis on the road he declared "don't go down there any further; you will be killed and you will lose your train and be blown to hell." Davis was acquitted, but resigned his commission about a week after the May 15th trial. [NARA, Record Group 153, file ii850]

saved their trains, had fled; and Williamsburg was occupied without opposition. Seven cannon had been abandoned, because the horses could not extricate them from the mud; and shot and shell were scattered in the road. The guns had been spiked; dishes, pans, kettles, and Dutch ovens, were demolished in the camps and redoubts; and packs of playing-cards had been thrown into the streets of the town. The soldiers visited every portion of the field to search for wounded or dead comrades, and witnessed without emotion scenes which lost their horror on account of their frequency; and, in many cases in which the death had been instantaneous, the attitude of the slain indicated the last action of their lives. The adjutant of a New Jersey regiment was shot while he was resting upon one knee, and glancing towards the advancing line of the enemy through a field glass; but his limbs had not relaxed.[23] Many hands, in every motion of loading and firing, clutched with the firm hand of death the rammer, gun, or cartridge. I noticed that a number of the dead of the Union army had been mutilated by bowie-knives, made of large, coarse files which the rebels carried in their sheaths; and gashes disfigured the heads and faces. The brief experience of a single engagement satisfied men of the uselessness of revolvers or dirks for the purposes of war, and they disappeared from both armies. Nearly one hundred of the enemy, in one part of the woods, had been killed while they were lying upon the ground; and the bullets had penetrated their foreheads. Some, who had lived a few hours after they were wounded, grasped photographs or letters, upon which their dying eyes rested when the thoughts of cherished and happy homes banished suffering. Some of the cartridges used by

------

23 The the only officer killed at Williamsburg matching Blake' description was Lt. Aaron Wilks, 6th New Jersey. This regiment also lost their commander that day, Lt. Col. John R. Van Leer. [Shreve, 16 and *Register of Officers and Men of New Jersey in the Civil War*]

the foe contained no bullet, but consisted of twelve buckshot. The pockets of friend and enemy had been turned inside out by the army thieves mentioned in the preceding pages; and the buttons of the uniforms of every traitor had been removed.[24] "I wish there was a battle every week," one of those miscreants remarked, in speaking of the amount he had stolen.

In wandering over the field, a corporal found in the pocket of a rebel a piece of tobacco upon which the blood had been coagulated, and the professional army thief had not touched it; but he washed the article, took a "chaw," and reserved the rest for future consumption. Head-boards were erected over the graves of the Union soldiers by their comrades, and the corpses of the enemy were buried by fatigue-parties; and the spot upon which they died was their last resting-place upon the earth.[25] Details interred the horses, large numbers of which had been killed. The woods and abattis upon the battle-field were burned for sanitary reasons; and the unexploded shells which had been thrown during the conflict were continually bursting in the flames. A surgeon ordered the pioneers to dig the grave for an unconscious and wounded rebel, who was supposed to be dying; but the general sent him to the hospital, and he lived.

Before the regiment had received orders to leave the field upon the 6th, an incident came under my observation which illustrates the difference between an officer of substance and one of show, and the wide contrast between a hero and a shirk. A brigadier-general of artillery, dressed in spotless apparel and

---

24 Confederate uniforms, when they could be obtained at all, were shipped without buttons. Recipients were expected to reuse the buttons from their old uniform.

25 "Many were buried on the spot, but most of those belonging to the Jersey and Excelsior Brigades were collected together and buried in long trenches. This could not well be done with the dead of our brigade, as they were scattered upon all parts of the field." [Haynes, 58-59]

white gloves, who, during the fighting of the previous day, was standing upon the ramparts of Yorktown, and watching the ripples that marked the wake of the transports when they steamed up the York, halted his horse, in the morning, near a battery which had taken an active part in the battle; and, as a matter of course, horses, guns, and men were covered with mud. He rebuked its commander, who still suffered from his fatiguing labors, for the dirty appearance of the artillery, and asked, in that arrogant tone of authority which characterizes many worthless officers, "Is that battery in a fit condition to move upon the enemy?"

"Yes, it is," he replied.

"Are you a regular? Do you say 'Yes' to me?"

"Yes, I said my battery was ready for service."

"Say, 'Yes, sir,'" he rejoined, and placed great emphasis upon the term "sir."

"Yes, sir," the lieutenant repeated with a salute; and the dignitary,[26] with his silver star and glistening gilt buttons, rode away. He was the type of large class of regular and staff officers, who always regarded external show, but never said anything about the services of a battalion; and the omission to use the word "sir," or a formal salute, was a greater blemish upon the record of a subaltern than the atrocious crimes of drunkenness, cowardice, or treason.

The reader has already observed that the inexplicable blunder of Gen. McDowell at the first action of Bull Run was committed upon a grander scale at Williamsburg, and four-fifths of the Army of the Potomac were non-combatants; and the

---

26 The general in "spotless apparel" was Brig. Gen. William F. Barry, who took a desk job in Washington after the Peninsula Campaign. By the middle of 1863, the muddy Capt. Osborn and Maj. Wainwright had become the Chiefs of Artillery for the XI and I Corps.

division was saved from destruction by the ability and commanding presence of Gen. Hooker.[27] In his concise report of this battle, he justly writes the following sentence: "History will not be believed when it is told that the noble officers and men of my division were permitted to carry on this unequal struggle from morning until night, unaided, in the presence of more than 30,000 of their countrymen with arms in their hands! Nevertheless, it is true."[28] A few shells, hurled from guns of a powerful caliber, wounded some of these soldiers who were near the field; and one general issued an address to his brigade after the conflict, and thanked the regiments for their courage and patience under fire, although they had not discharged a cartridge in the fight. This document was published in the newspapers of his State to advance his political interests.[29] The troops that delayed the regiment upon the Yorktown Road, at the Half-way House, like those that obstructed the Warrenton Turnpike at Centreville,[30] preceded in the order of march, but took no active part in the action which followed.

The total loss of [Hooker's] division in battle was 1,575; of which there were 338 killed, 902 wounded, and 335 prisoners and missing.[31] When these figures are compared, it will be seen

27 The timely arrival of Kearny's division probably had more to do with it than Hooker's commanding presence. Hooker precipitated the battle on is own initiative, without consulting III Corps commander Heintzelman.

28 Blake is quoting Hooker's report. [OR XI/1, 468] See Appendix: "Hooker at Williamsburg."

29 Probably Brig. Gen. Charles Devens, whose brigade suffered one killed and two wounded while in reserve of Peck's brigade. Devens was a Harvard law graduate who had been a state senator. Post-war he served the state as a superior and supreme court justice.

30 At the commencement of the Union flank march at First Bull Run.

31 Hooker's division did the bulk of the fighting at Williamsburg, and was outnumbered by the Confederate forces facing it. The official tally of losses in the first brigade were modest compared to the other two, which

that the number of wounded was small; and the fact shows the deadly proximity of the soldiers to their enemies upon the field. The total loss of Gen. Hancock's command was 31. I mention the last item because a disposition has been shown by certain parties to magnify the action of Gen. Hancock in this engagement, and deprive Gen. Hooker of that credit which he had so well merited. Neither Gen. Hancock, nor the officers and men of his brigade, ever made any claim of this character, but took the opposite ground, and refused to accept the meed of praise which they deserved.[32]

The regiment encamped in the suburbs of Williamsburg;[33] and the comfort of most of the men was increased by living in tents which the rebels had been obliged to abandon in their hasty retreat. The crumbling statue of Baron de Botetourts; who had been "Governor-General of the Colony and Dominion of Virginia," was typical of the decay of this portion of the State; for the ancient city had lost its former importance, and was now celebrated as the seat of the College of William and Mary, in which some of the most eminent statesmen of the United States had been educated.[34] All the desolations of war, the legitimate results of the Rebellion, were visible throughout its limits; and the public buildings, halls, churches, and many dwelling-houses, were filled with the wounded of both armies. The yellow flags,

---

bore the brunt of the Confederate flank attack.

1st Brigade (Grover's)	252
2nd Brigade (Excelsior)	822
3rd Brigade (Jersey)	526

32 "Meed" is an archaic word meaning "just reward." Perhaps Blake intended to say "...praise which they *did not* deserve."

33 Grover's Brigade was assigned to provost duty in Williamsburg through May 15.

34 Including Presidents George Washington, Thomas Jefferson, James Monroe, and John Tyler.

which indicated the rebel hospitals (red was the color of the Union hospital-flag), waved in every district. The recitation-rooms of the college; the aisles of churches, from which the pews had been removed; and the marble slabs in the grave-yards that adjoined them, — were stained with the blood of mangled soldiers. The people, with few exceptions, were traitors, who had always encouraged those that murdered the forces that upheld the National Government; and the closed and empty stores, the absence of the able-bodied white men, the scowls of the women and children, and the delighted faces of the negroes, were perceived by the most casual observer. When a squad of Federal prisoners arrived during the afternoon of the battle, the rejoicing populace loudly cheered over the victory which they considered won; and some, who armed themselves with axes and clubs, expressed an intention to kill the wounded upon the field as soon as the [Union] army retreated to Yorktown. The falsehoods of Northern and Southern rebels had been accepted by them as facts; and one-half of the population, fearing that the troops would commit the grossest outrages, fled to Richmond as destitute and terror-stricken as the settlers upon the frontier when the torch and scalping-knife of the savage commenced the work of destruction. No wagons or horses were seen in the streets or stables of the town; and the slaves lived in the mansions of the fugitives, and enjoyed the privileges of freedom. Persons who had refused to flee, and ignorant women who had been left helpless by their male relatives in Johnston's army, bolted the doors and closed the blinds of their domiciles, and shuddered when they thought of the "monsters of Lincoln," who had a "heart of iron." The conduct of the Union soldiers, after the occupation of the place, which was humane and just, as it always has been upon every occasion, convinced them of the groundlessness of their apprehensions; and the shutters were once more opened to admit the rays of sunlight into their cheerless homes.

They stated, in conversation, that Judge Bowden was the sole Union man in the town; and he remarked to the troops when they took possession of it, "The sound of the first volley of musketry was music to my ears." This patriotic citizen subsequently represented the loyal people of Virginia in the Senate of the United States.[35] A detail from the regiment guarded four hundred prisoners, who excitedly discussed the alleged demerits of their respective States; and it was necessary for the sentinels to interpose in several instances to prevent encounters between them. The system originated by Gen. McDowell of rigidly protecting rebel property, which was one of the leading ideas of the commanding general, began at this time; and the force of Gen. [Andrew] Porter, Provost Marshall of the army, upon the staff of Gen. McClellan, was constantly employed upon the Peninsula in performing this odious task. While the soldiers were obliged to obtain passes to escape the custody of the patrols, the rebel surgeons and hospital nurses, who came inside of the lines under a flag of truce, traveled in every direction unmolested, and were allowed to use their negro servants as property.

The women, destitute of every trait that constitutes the lady, who had been so cowardly in the time of imaginary dangers, took advantage of the uniform courtesy of the "Yankees," whom they despised and hated, and haughtily walked in the streets with their "niggers," who carried dishes and baskets of luxuries and food for "missus," who distributed them among the sick and wounded

---

35 Lemuel Jackson Bowden (1815-1864) was a prominent lawyer in Williamsburg whose Unionist opinions made him the object of scorn about town. Union forces held Williamsburg for the remainder of the war, installing Bowden as mayor. [Leech, 283] Bowden was elected U.S. Senator in 1863 by the Unionist shadow government of Virginia.

rebels. They compressed their dresses whenever they met an officer or enlisted man, so that the garment would not touch the persons they passed. They pulled their hats over their faces to preclude scrutiny: but those precautions were useless; for their cadaverous features and lank forms were sometimes seen; and all were satisfied that the Southern beauties, about whom so much has been written, did not reside in Williamsburg. They gladly paraded through the mud and filth of the street to avoid a squad of men upon the sidewalks. When two young rebel females were walking by some soldiers, one of them suddenly screamed like an affected boarding-school miss who beholds the horrible form of a dreadfully shocking beetle, or an awfully distressing toad. "Oh! Oh! What have you done? Your skirt touched a Yankee!" A group of these sympathizers gathered around the bed of a sufferer in the hospital who needed rest to recover, and persisted in talking together, and striving to make him more comfortable, until they produced an unnatural excitement, which was speedily followed by death. "Dear hero, I must assist you." "Noble soldier, what shall I help you do?" or "You deserve everything we can give," were the sentiments which they generally expressed, in a theatrical style and tone; and many wretches were thus killed by the ill-timed conduct of these well-meaning friends.

Chapter V

The March, and Fair Oaks

THE DISTANCE from Williamsburg to Richmond is fifty miles: but the advance of the troops was extremely slow; and upon May 30, twenty-five days after the battle, the division, forming the left wing of the army, was encamped at Poplar Hill,[1] near Oak-Bottom Swamp, which was thirteen miles from the rebel capital. The dates of the bivouacs during the period would be uninteresting; and the minutes from day to day speak continually of storms and miserable roads, deserted houses and farms, thick forests and a scanty population: and the only objects that relieved the monotony of the march were the "White House,"[2] and the church in which Washington was married.[3] A bridge one hundred and twenty feet in length had been constructed at Bottom's Bridge, over the Chickahominy, which flowed through the swamps and bottom land, and enlarged its banks in the season of freshets until it was half a mile in width. The perusal of signboards might be classed as dry reading; but every regiment in the division was convulsed with laughter when it marched by a board nailed to a tree at the fork of the road, upon which was painted, "Richmond, 17 miles."

Horses, sheep, houses and forage, in every part of the country, were heavily guarded by the command of Gen. A.

---

1   Grover's Brigade left Williamsburg on May 16, taking fourteen days to arrive at Poplar Hill, the high ground overlooking the principle crossing of White Oak Swamp.

2   Home of Robert E. Lee's son W. H. F. Lee at the head of navigation of the Pamunkey River. White House Landing was the Union army supply base until the end of June, at which time the house was burned to the ground.

3   St. Peter's Church, near White House Landing, where Washington married Martha Custis Lee.

Porter;[4] and, so zealous was this officer in the performance of his duty in this respect, soldiers [as guards] were put upon property that was not within the limits of the Union lines, and details from different regiments were required when this large force was inadequate. The wives and daughters of the owners, who were in Johnston's army, insulted in every way the national troops, and rewarded those that protected their estates by acting as spies, and informing the guerrillas of the opportune moments in which they could capture or murder them. Officers and soldiers were sometimes driven from wells by the sentinels, and compelled to drink the water they could find in the brooks and springs in the fields; and colonels and commanders of batteries were not allowed to take forage for their suffering animals when it was impossible to procure it from the depot of supplies. The spirit of the instructions which the guards received is exhibited in the language of an infamous order which Gen. McDowell issued at Fredericksburg, "to place a sentinel upon every panel of fence," "if it should be necessary," to prevent the men from taking the rails to promote their comfort.[5] Not a solitary rebel was willfully injured by generals who seemed to forget that treason was an offense, and that, as Lord Belhaven[6] declared in the Scotch House of Lords, "patricide is a greater crime than parricide, all the world over." This cringing forbearance towards the enemies of the country disheartened the troops; was viewed with contempt by the rebels, and considered proof of weakness, although they admitted that their property was better protected by

4 Provost Marshall of the army, who was attempting to protect civilian property.

5 While occupying Fredericksburg, April-May 1862, McDowell ordered the protection of civilian property, having witnessed extensive destruction by his army during the Bull Run campaign.

6 Alluding to a 1706 speech by Scottish Lord Belhaven concerning the proposed Act of Union between Scotland and England.

the army of Gen. McClellan than by that of Johnston. The colonel of one regiment, who was an M.C.,[7] often submitted to his own judgment certain orders that he received, before he complied with them, and this policy was very obnoxious to him. He was once commanded to keep his troops under arms two hours for a trivial infraction of the rules of discipline in refusing to report the names of some so-called offenders.

"I sha'n't do it," he remarked to the brigadier-general, as he walked to and fro, and whistled when he was not smoking.

"Did you understand that this was my order?" inquired the general.

"I don't care for you: we make fellows like you in Congress," the M.C. replied; and the regiment did not suffer punishment.[8]

The stupid and inhuman treatment of the negro bondmen at the first Bull Run was blindly adopted by Gen. Halleck in the Western Department and by the commander of the army in Virginia [Maj. Gen. John Pope].[9] The slaves possessed the most valuable local information concerning fords, roads, the divisions of the foe, and the forts that environed Richmond; and this could be obtained by persons with acute minds and a knowledge of human nature: but the so-called Orleans princes,[10] who had a partial charge of the secret-service department, were wholly

------

7 Col. Gilman Marston, 2nd New Hampshire, was a Member of Congress from 1859-67.

8 The shocked brigadier general was Cuvier Grover for whom the rank and file had a high regard, despite Marston's disdain. [Haynes, 117]

9 Blake refers to the policy of turning away fugitive slaves if they tried to enter Union lines seeking refuge.

10 Louis-Phillipe d'Orleans, *comte de Paris* and pretender to the French throne, and his brother Robert, *duc de Chartres*, served on McClellan's staff analyzing intelligence reports. They were accompanied by their uncle, the Prince de Joinville.

unfitted for this important office. The sad results of regarding the tales of negroes as valueless, and employing men who imagined that they were born to be kings, are well known. The force of the enemy was always magnified at headquarters; and the army that Lee directed during the "Seven-days' Battle" was asserted to be one hundred and eighty or two hundred thousand soldiers.[11] The generals who expelled slaves from their lines in public orders, and rejected their facts, displayed the same depth of ignorance as the despot who cimeters the heads of his couriers if they bring the unwelcome news of reverses. Some extracts from Shakespeare[12] are so apposite, that I quote them:

> Ædile: *There is a slave, whom we have put in prison,*
> *Reports—*
>
> Brutus: *Go see this rumorer whipp'd. It cannot be—*
>
> Menenius: *But reason with the fellow,*
> *Before you punish him, where he heard this;*
> *Lest you shall chance to whip your information,*
> *And beat the messenger who bids beware*
> *Of what is to be dreaded.*

The wise advice of Menenius represents the policy of the present mode of gaining intelligence in fighting the Rebellion.

On [the night of] May 30 [and 31] fell one of the most severe storms of the year, which inundated the swamps, nearly severed the communications between the forces upon the north and south banks of the Chickahominy, and washed away a space of the road that bounded the camp, about twenty feet square, to the depth of a yard. The rain continued at intervals during the

---

11 These figures reflect the estimates of General McClellan. The actual number was closer to 100,000.

12 *Coriolanus*, Act IV, Scene vi, in which Roman statesman Menenius Agrippa advises Brutus of the more prudent policy.

next day [June 1]; and at two, P.M., Gen. Hooker marched five miles to the battle of Fair Oaks with the second and third brigades;[13] while Gen. Grover's command held the position, and remained in the line in readiness for the onset which was momentarily expected. "Hold your ground at all hazards!" were the only orders that were received. Upon this afternoon, and morning of June 1, the men listened to the dull reverberations of the distant conflict with intense anxiety, which was relieved by the arrival of dispatches that contained accounts of the decisive repulse. The brigade joined the troops at Fair Oaks upon the 3d; and Gen. Casey's division, which had suffered a severe loss in the battle, and was destitute of tents, clothing, and cooking-utensils, that had fallen into the hands of the enemy, occupied Poplar Hill: so that the two commands exchanged their positions in the line.

The regiment relieved one of the Excelsior Brigade in the midst of a drenching rain and the darkness of the night of June 3, and performed its tour of picket-duty for twenty-four hours. Although the defeat of Johnston had been complete in this terrible contest,[14] and the authorities in Richmond expected an immediate pursuit of their demoralized forces, the extent of their losses was so slightly understood, that the pickets were always urged to be vigilant, because an attack by the enemy was hourly anticipated at this point; and the men stood in line of battle before twilight. The field was visible in the morning to the eyes of the soldiers, who beheld one of the most ghastly spectacles that has ever been witnessed. Scores of horses, and the swollen

---

13 The two brigades of Hooker were sent in support of Casey's division (3/IV) south of the Williamsburg Road, which was struggling to repel the attack of Huger.

14 Both sides claimed victory at Fair Oaks. The Union occupied the battle field, but Johnston had not been defeated.

and black corpses of hundreds of rebels, were stretched upon the ground, and in spots lay in groups, that showed a fearful waste of life; and myriads of maggots were feasting upon the putrid forms, and swarmed upon the earth, so that it was difficult to walk without crushing them beneath the feet. Many soldiers, in the obscurity of the night, had slept side by side with the bodies of the slain, supposing that they were comrades; and the loathsome worms entered their haversacks, and crept upon their blankets and overcoats. Some, who had complained about a foot or boot that interfered with their personal comfort, of the form of a person over whom they had stumbled when groping the way to their posts, were amazed to discover that a corpse had been the subject of their oaths. Others, who collected wood to cook coffee and build light fires, found that they had taken the rude headboards which the rebels had placed over the graves of those they had buried. The stench was continually aggravated in its intensity, until an unyielding military necessity was the only power that made it endurable. The fragments of shells, the *débris* of the camps of Gen. Casey's division, and the bivouacs of the enemy, were scattered upon the battle-field; but every article had been destroyed. Every object showed the marks of the great struggle; many camp-kettles had been pierced by bullets; and a Sibley tent,[15] that stood in the midst of a shower of lead, displayed two hundred and forty-six holes. In the forest, small trees an inch and a half in diameter were cut in nineteen or twenty places, and limbs upon the large oaks had been splintered by shells. A cannon-ball, three inches in diameter, would sever a branch that was five inches thick; but, when it passed through the massive trunk, the elastic fibers closed up the aperture, so that I could not insert my sword to the depth of an inch. A portion of

---

15 Cone-shaped canvas tent with a central pole and low walls; it could sleep
    up to a dozen men.

the dead were buried by the foe during the occupation of the camps upon May 31; while the largest number were plundering knapsacks and the tents of the quartermaster and commissary department, and threw aside their dirty and ragged uniforms, and wore the comfortable garments and underclothing of the Federal soldiers. A few barrels of whiskey fell into the hands and mouths of the [rebel] victors, and rendered many of them unfit for the conflict which was resumed upon the following morning. One of these greedy privates, who had succeeded in arraying himself in three pairs of pants, was killed while he was putting on another pair which he held in his hands.

The topography of the battle-field, which resembled in its general features that of Yorktown and Williamsburg, was a swampy plain covered with woods, with the exception of a cleared tract of land upon the Williamsburg Road that was locally known as "Seven Pines." Fair Oaks, which was a station upon the York-River Railroad, about half a mile from this point, was the name given to the battle by the national forces; but the Richmond newspapers, and the commander of the enemy, called it "Seven Pines." The mud, which forms a leading subject in the history of this war, prevented both sides from using their artillery to a vast extent, and the contest became one between the infantry. Many generals expressed the opinion that the army could have taken the beleaguered capital at this time; and Gen. Hooker remarked in a very sanguine tone, when speaking of the matter, "Phil Kearney and I could have gone into Richmond."

The division was encamped upon the field in the rear of Seven Pines [from June 3] until June 30; and, during this period, its history comprised a record of labor upon breastworks; and, once in three days, Grover's brigade relieved the Jersey brigade, which relieved that of Sickles, and performed picket-duty for twenty-four hours. Graves were visible in every direction, after the horses had been burned and the dead were buried: and, when the line was advanced, some were seen in the swamp, standing in the posture in which they were killed; and, so rapidly had they decomposed in three weeks, there was no flesh upon the skulls, which had partially bleached. Rifle-pits and redoubts were constructed by the division; and acres of the forest were felled to obstruct the foe, and allow the artillery to have a point-blank range upon the advance. The pickets were stationed in the edge of this abattis, and supported by the reserve that was posted behind the extensive fortifications. The hostile lines were engaged in an incessant skirmish at times, and the cannon frequently threw a few rounds into the woods in which the enemy was concealed. The expectation of an attack was so strong, that the troops were always formed for battle; and the regiment was called to arms upon one day eleven times. The bugle at brigade headquarters sounded the order to "fall in," and the soldiers rushed to their stacks whenever the firing in front was unusually active. The rebels made a reconnoissance during a storm, and delivered a volley in a peal of thunder: but the practiced ears of the men were not deceived; and they left their tents and double-quicked to the line, upon which they formed before the call was blown. Shells were often thrown into the camp; and one of them, during a cannonade, entered one side of a tent in the company while the inmate was coming out of the entrance. There was much useless firing upon picket; and, while some regiments were comparatively quiet, others would shoot at random, and keep the supports and camps in a state of constant alarm and preparation. An emaciated rebel came into the lines one morning, and confessed that he ran away from his regiment in the battle, hid in the woods, and did not wish to return to it, because he thought he would be shot. The pickets fired at him whenever he approached their posts; and he remained in his place until he was compelled to escape from death by starvation, and run the gauntlet of bullets. The enemy tried to ascertain the position of the sentinels by the use of dogs and hounds, which ran through the forest, and barked when they saw any person; the sharpshooters always killed them: and sometimes the dismal howl of a wounded cur limping to his master interrupted the quietness of the hour. The same causes that existed at Yorktown, again affected the health of the men; and the water, which was tainted by the decomposition of the bodies that had been interred near the camps, increased the long list of the sick. A ration of whiskey was daily issued to the troops to avert the malaria of the swamp; and this remedy was more enervating than the disease. I noticed, as a singular circumstance, that there was not a single case of the "Chickahominy fever"[16] in the small number of those that refused to obey this order.[17]

Gen. Stuart made his celebrated raid in the rear of the army upon the night of the 13th; and, although its material results were unimportant, it frightened the sutlers and non-combatants, and proved the inefficiency of the Union cavalry.[18] Here, as at Yorktown, the heavy burdens of war were placed upon the shoulders of a part, while some performed no real service; and there were regiments that had never discharged a musket in

---

16 A combination of mosquito-borne malaria and typhoid fever caused by the unsanitary water supply.

17 During its weeks-long stay near Fair Oaks, another officer of the 11th ran afoul of the military justice system: Capt. Selden Page was brought up on charges for failing to go on picket duty. He apparently was overcome by the noxious conditions described by Blake, but failed to get a medical certificate. He was convicted and sentenced to forfeit a month's pay. Page resigned on July 2.

18 Stuart's reconnaissance mission behind the Union army with 1,200 cavalrymen proceeded unopposed and turned into the "Ride around McClellan." Union cavalry at the time was dispersed throughout the army and unable to concentrate against Stuart.

battle, or labored upon a fort, or served a tour of picket-duty in front. When the company was deployed upon the outposts, upon one occasion, there was so much infantry firing about three miles in the rear of Fair Oaks, that it was supposed the enemy had made an attack; and the pickets were anxious about the result, until the commander of the brigade learned that the regiments of a division were drilling as skirmishers, and using blank cartridges.

The troops that were constantly employed were annoyed by noxious vermin, that lived in every resting-place in the front: generals and privates, however vigilant, were defiled; and every article of clothing was scrutinized by the men when they were relieved from picket. Many regiments in the army of the Potomac had been depleted by sickness and desertion (for the loss in battle had been limited to a small fraction), and those which re-enforced it appeared as large as brigades.[19] All were encouraged by the arrival of these troops, who sometimes mentioned, as a trying hardship, the fact that they had had "no soft bread for two days." The surgeons in certain hospitals that were located in the North began a course of conduct at this time that was long continued, and placed upon their sick-rolls the names of deserters and cowards who were feigning disease when the country required their services in the field.[20] There was scarcely a regiment upon the Peninsula that did not lose at least one hundred able-bodied men by the connivance of these medical miscreants. Eight soldiers deserted from the company at Williamsburg, and were transported to a notorious hospital in

---

19 In mid-June the full-strength 16th Massachusetts joined Grover's brigade. It had served garrison duty at Fort Monroe since September 1861 and had just participated in the occupation of Norfolk, Virginia.

20 This is Blake's first remark on the hospital system's harboring of malingerers, which he expands on in Chapter XVIII.

Rhode Island,[21] in which they were retained and sheltered by the officers, who were repeatedly notified that they were skulkers; and none of them ever rejoined the regiment, although their term of enlistment did not expire for two years. One of the number was employed by some of the surgeons and nurses to repair and manufacture jewelry, while the rest were engaged in similar avocations; and none of them were treated as patients. In striking contrast with such shameful conduct was the noble action of men who returned to their commands before their wounds were healed, and those who were excused from duty on account of sickness, but left their beds, and walked with great difficulty to the front with their muskets. I have heard officers tell them that their presence was not required; but the same answer was invariably made: "I could not stay in the rear when I thought the regiment was fighting."

The supplies were brought from the stations upon the York-River Railroad in wagons which were parked in the rear of the army at night to avoid the risks of capture.[22] The whistles of locomotives, and the rumbling of the trains of cars, blended strangely with the shrieking shells, and suggested thoughts that were as conflicting as war and peace. A sentinel could discern the steeples of Richmond, which was six miles from this point, from the top of a tall tree near the front, called the "lookout." A company of Frenchmen, that belonged to a regiment that had been inveigled at Williamsburg by the white color that has been described in the sketch of that battle, was posted upon the picket-line, and fired upon a flag of truce which they supposed to be

---

21 Lovell General Hospital, Portsmouth, Rhode Island, opened in July 1862. Among its patients were Union court-martial convicts and Confederate prisoners of war, necessitating a detachment of state militia posted as guards.

22 Five locomotives and dozens of freight wagons had been transported by sea to White House Landing, the terminus of the York River RR.

another artifice of a dishonorable enemy.[23] While a soldier, who had been sleeping, was walking a short distance, two pieces of a shell penetrated his blanket which was spread upon the ground; and with the remark, "Did you see that strike? it was a lucky escape," he slept in the same place. An austere colonel of a New Jersey regiment sat upon a stump during one of the severe storms that often fell: and a drummer, who observed that he had no shelter, brought him a rubber blanket, and said, "Colonel, take my blanket; you will get wet."

"Clear out, you scoundrel!" was the gruff reply.

"It is raining hard, and you will need it."

"Go to your post at once."

No further remarks passed between them, and the conversation, like many others, is quoted to illustrate the eccentricities of the commander; but a commission as first lieutenant was given to the musician at the end of a fortnight.[24] Gamblers to the number of three hundred, equipped with the implements of their nefarious work, — dice, props, and cards, — assembled near the regiment after the army had been "paid off," and disregarded the [cannon-]balls that sometimes ploughed the field, until a blast of the bugle summoned them to disperse and enter the line of battle.

The [Third] corps was formed [for battle] upon June 25;[25] and the regiment, with others, advanced at seven, A.M., through the woods and swamp, that was one-third of a mile in width, and

---

23 55th New York of Peck's brigade (3/1/IV) consisted predominantly of French immigrants, but their involvement with a white flag incident at Williamsburg has not been identified.

24 The "austere colonel" was Samuel Starr, 5th New Jersey, who was known as a strict disciplinarian with an explosive temper.

25 The first of the Seven Days. McClellan launched a limited offensive up the Williamsburg Road to secure dry ground for his siege artillery around Old Tavern.

halted near the open field, in which a burnt chimney stood.[26] The pools of stagnant water, the stumps, and the thickets, continually threw the line into disorder, and, together with the inability to see the force of the enemy that confronted them, caused a slow movement.[27] Every object that looked like a rebel received a bullet: the [rebel] pickets, leaving their rations and blankets, hastily fled; and two sharpshooters, perched in the tops of trees, were captured before they could escape. The [Excelsior Brigade] regiments upon the right of the brigade met the [rebel] reserve, which was re-enforced by a detachment. Their ranks were shattered; small squads, which increased in number, began to leave; and at length all were flying to their works, amidst the excited cries of the men [of the 11th Massachusetts], who said, "They are running!" or, "Look at that fellow tumble!" and the [rebel] troops that [had] double-quicked to the front during the fight fled from it with greater speed.[28] The regiment remained undisturbed[29] in its place in the centre of the brigade. The firing was very active in the afternoon, when Gen. Kearney's division upon the left was attacked,[30] and the right of Gen. Hooker's division was advanced; but the contest had ended at four one-

---

26 1st and 11th Massachusetts were thrown forward as skirmishers for Grover's Brigade, on the right and left, respectively.

27 The Excelsior Brigade's advance on the right of Grover was slowed by boggy terrain, leaving the 1st Massachusetts' right somewhat exposed.

28 Blake's convoluted description of the action leaves out the repulse of the Excelsior Brigade, halting the advance.

29 McClellan stopped the attack by signal telegraph at 11 o'clock, but changed his mind after coming to the front to see the situation. Hooker wrote: "suddenly the major-general commanding the army, appearing on the field and learning the state of affairs, gave directions for me to resume and finish the duty assigned to me the night previous." [OR XI/2, 109]

30 This was a principal point of counter-attack by Wright's Georgia brigade of Huger's division. For a brief time, Georgians in red Zouave uniforms were mistaken for Union troops. The fighting petered out at sunset.

half, P.M., and the troops held the ground which they had been ordered to take.[31] The aides could not move through the forest upon their horses; and commands were passed from company to company along the line by shouting, "Keep a sharp lookout to the extreme left!" "Tell Col. _____ to report to the general upon the right!" and others of similar character.

I was placed in charge of the skirmishers who were stationed in advance of the regiment to prevent surprise, and heard the conversations of the enemy's pickets, who were separated from them by a morass that was covered with thick woods.[32] Two of them had a quarrel about Gen. Kearney, who they styled "the one-armed devil:" one swore that it was the left, and the other was certain that the right limb had been amputated.[33] The solitary bass drum that was constantly beaten in the rebel camps during the siege was unusually distinct in its notes;[34] and an [enemy] officer in front shouted orders, which were repeated by three or four voices, for the period of four hours. "Deploy upon the right of the Williamsburg Road, and don't get into seven or eight ranks;" "Advance the skirmishers cautiously up the paths;" "Why don't those men move forward?" were the commands that excited the vigilance of the soldiers. The loud tone in which they were uttered, the absence of sounds that would be caused by the tramp of moving columns, and the subsequent conduct of the foe,

---

31 The 11th suffered only eighteen wounded at Oak Grove, none killed. Blaisdell wrote: "I am most thankful for the almost miraculous escape of my men being injured by the very heavy and continuous fire of the enemy." [OR XI/2, 126]

32 Blake later noted that he "may have been as near Richmond in 1862 as any Union soldier who was not a prisoner." [Familiar Quotations]

33 At the Battle of Churubusco, Mexico (1847) Kearny had been wounded in the left arm, resulting in amputation.

34 Drums were used in battle to transmit signals because their punctuated sounds traveled further than voice commands.

proved that it was a stratagem to mislead the [Union] commander, and induce him to draw re-enforcements from other points; while a fierce and unsuccessful attack was made upon Gen. Kearney's division, which was posted upon the left. The members of the regiment clustered around the roots of trees in the night [June 25-26], and sat upon clumps of earth, because they were obliged to stand in water that was knee-deep when the line of battle was formed. "The endless groan" of the [Confederate] wounded, and the rattling wheels of the wagons that conveyed them to Richmond, alone disturbed the stillness that reigned in camp. A [Federal] regiment was marching to relieve the troops at the front at midnight:[35] the shovels, canteens, and equipments which they carried were constantly clashing together, and the sounds alarmed the enemy [pickets] that fired at them, and revealed by the flashes the positions of the pickets. A flame that lighted up the forest for an instant darted from the smooth-bore muskets [of the 11th], to which there was no response; and a rebel sentinel swore at his companion, "Don't fire again, you fool! you will draw upon us another volley like that." The division returned to the rifle-pits: the first of the "seven-days' battle" before Richmond had been fought upon the left with a successful result; and the position that had been gained was entrenched.

A regiment composed of different nations,[36] which was well known on account of its cowardice, was upon outpost-duty, under the command of a foreign lieutenant-colonel, who excitedly exclaimed to his men, "Cover yourselves mit a stump!

---

35 Blaisdell reported that the 11th was "relieved by the One-hundred-first Regiment of New York Volunteers, at 2 o'clock this morning [June 26]." [OR XI/2, 126] The 101st New York was from Birney's brigade of Kearny's division.

36 The "regiment composed of different nations" at Oak Grove has not been identified.

cover yourselves mit a stump!" They were in the rear of the advanced ground which had been conquered: volleys were fired into the woods whenever a bullet, passing from the front, whistled over their heads; and many of them fled from their posts during the night. The whole detachment, with the commander in the advance, rushed toward the works on the morning of the 26th; and the colonel [Blaisdell] at once deployed a company to stop the fugitives, and gave this instruction: "Use your bayonets upon the cowardly scoundrels: they are not worth the powder to blow them to hell!" The captain promptly halted the lieutenant-colonel by the use of physical force, and ordered him to return to his post of duty.

"I out-rank you, capitain," he said as he displayed his shoulder-straps, and refused to move to the front.

"I don't care for your rank: you must go back to that picket-line."

"My mens run away, and leave me: I no go back."

"You ran away first, and they all followed you. You can't see one of your regiment in front of you." The determined conduct of the line, who is now a field-officer in the service, and the menace of physical violence, intimidated the poltroon, who sullenly skulked to his command. A month after this shameful occurrence, I saw with amazement, in a New-York illustrated paper (it was not Harper's), an engraving, in which this regiment appeared to be capturing a battery, and driving the brigades of the enemy; while the lieutenant-colonel, mounted upon his war-horse at the head of it, was cutting "horrid circles" with his sword.[37] Through the untiring exertions of certain officers, when Gen. Meade commanded the Army of the Potomac, he was promoted to the rank of brigadier-general for "gallant conduct upon the field."

---

37 Not found. Our search of *Frank Leslie's Illustrated Newspaper* for the last half of 1862 turned up no engraving matching Blake's description.

The perusers of the foregoing facts can readily imagine the nature of his valiant services. He is upon detached duty in a Northern city at the present time, and inspects the harbor defenses, or acts as pall-bearer at the funeral of officers of high rank who have died the death of heroes.[38]

The enemy made feints upon the entire line: the division was posted behind the breastworks, in readiness to meet the onset; but the grand [Confederate] assault was made upon Mechanicsville on the 26th, and Gaines Mill on the 27th. The troops of the [Third] corps were withdrawn upon the 28th from the position which had been taken upon the 25th; the army made preparations for the retreat during the night; and officers who had seen the smoke of the burning bridge[39] in the afternoon, and knew that the communications with the "White House" had been severed,[40] refrained from giving this information to their commands. The [Confederate] balloon had constantly made reconnoissances; and one rose for the first and only time above Richmond upon this day, and remained in the air a few minutes.[41] All the stores that could not be transported in the wagons were destroyed early in the morning of the 29th by details, who broke rifles, bayoneted canteens and kettles, and slashed tents and clothing, but burned nothing, because the fires would excite suspicion. Barrels of sugar, vinegar, and whiskey irrigated the soil of the camps; and some soldiers, who were unable to restrain their appetites,

---

38 Despite Blake's lengthy exposé, the miscreant "foreign lieutenant colonel" of June 26th escaped positive identification.

39 Bottom's Bridge over the Chickahominy.

40 After the battle of Gaines Mill, McClellan decided to move the entire army south of the Chickahominy and establish a supply base on the James River at Harrison's Landing.

41 The only source of hydrogen gas was the Richmond Gas Works. The inflated balloon would be attached to a train car and towed to its observation point.

stealthily drank the intoxicating liquor, were left upon the field, and captured by the enemy in a state of utter drunkenness. Gen. Hooker destroyed his personal baggage, and set an example of unselfish patriotism, which might have been followed by other commanders who encumbered the trains with their private goods, and cheerfully abandoned the property of the Government and that of the men and subalterns. The brigade retired from the scene of its labors after it relieved those upon picket; and I never beheld so many faces upon which was depicted such a deep feeling of gloom.

Chapter VI

Savage's Station, Glendale, and Malvern Hill[1]

THE MOVEMENTS of a vast army are slightly understood by the men who perform the fighting; and my knowledge of the positions held by the [III] corps during the retreat was obtained by noticing the reports of cannon when they were engaged with the enemy; and I sketch in this chapter some incidents in the action of a small body of troops. The company was posted, on the 29th [of June], at a house upon a hillock near the railroad; and a line of battle, which extended more than a mile towards the right, had been skillfully formed in the edge of the forest, in the rear of a cleared space of ground, upon which the batteries were planted, near Savage's Station.[2] The occasional report of a sharpshooter's rifle was sometimes heard, and many were sleeping under the peach and forest trees which shielded them from the hot rays of the sun. The artillery of the rebels opened without any warning, at eleven, A.M., from the woods upon the south side of the railroad, on this position with wonderful accuracy; and the first three shells that were fired penetrated the walls and partitions of the house,[3] and mangled those who were one-fourth of a mile in the rear, but did not injure the inmates,

---

1 Blake is referring to the large-scale reconnaissance conducted by Hooker's division in early August.

2 June 29th saw two similar actions along the York River RR: a brief clash at Peach Orchard (Allen's Farm) in the morning, and a more serious battle at Savage's Station (two miles back) in the afternoon and evening. Because of the similarities (Magruder attacking Sumner in both), regimental historians have recalled both as the Battle of Savage's Station. The report of General Grover [OR XI/2, 122] indicates that the brigade witnessed the morning Peach Orchard battle, positioned on the left of Sumner's (II) Corps.

3 Possibly the Allen house, half mile in front of Grover's battle line and south of the railroad.

comprising women, children, and a squad of soldiers. Upon a bed, and unable to move, was a sick woman, whose husband and sons were with those who were trying to murder her; and the shot and shell were not hurled into the ranks, but purposely aimed at this dwelling; and the cries of the helpless infant and the tears of the distracted mother were stifled by the explosions and shrill notes of the flying balls, until their batteries were silenced by Union cannon. An attack was afterwards made upon the brigade which was enveloped in the smoke of battle;[4] and joy filled every eye when the breeze gradually lifted the veil, and revealed the American flag that waved over the victors, who were still invisible. The division fell back from its first line in the afternoon;[5] moved with rapidity; crossed the White-Oak Swamp; and at ten, P.M., bivouacked in a field near the Charles City Road. Clouds of black smoke rose at certain points near the railroad stations, and immense amounts of clothing, provisions, and ordnance stores, were destroyed. Although the troops halted at Savage's Station, and many needed the garments, guards were

---

4   The 11th was the closest unit to Sumner's Corps. Grover reported that: "a strong picket from the Eleventh Massachusetts, under Major Tripp, was thrown out, covering the various approaches to our position. The enemy, however, did not move upon us in force, but directed his whole attack against General Sumner's position on our right, only throwing occasionally a few shell into the woods occupied by my command. Fortunately, though many shells fell within our limits, no one was killed, and but 2 men of the Eleventh Massachusetts Volunteers, who were on picket duty, were wounded." [OR XI/2, 123]

5   After about two hours of fighting at Allen's Farm, the entire Union force fell back toward Savage's Station. As the Union position there took shape, Heintzelman (III) unilaterally decided that with his corps, Sumner's (II), and Franklin's (VI), there were more men than "could be brought into action judiciously." [Ibid. 99] Thus, shortly after noon on the 29th, Heintzelman headed Kearny and Hooker south through White Oak Swamp. Sumner was very upset about this disappearance, but he and Franklin were able to hold off the Confederate attack and cross the White Oak Bridge later that night. McClellan was not present and had left his corps commanders to sort out the problems of coordination.

posted to prevent the soldiers from taking a blouse or coat; because the officer in charge had been ordered to burn them, and could not account for the property if it was worn by the men.

The brigade rested near the church at Glendale[6] upon the 30th [of June], and trains of wagons and herds of cattle — "beef on the hoof" — were continually passing over the road during the forenoon, while an active cannonading in the neighborhood of the bridge at White-Oak Swamp showed that the enemy was closely following the army. There was a panic at one time among the teamsters, who fled from their horses in the most cowardly manner; and the cavalry, with their drawn sabers, forced them to return to their seats and resume the reins. The divisions of Hill and Longstreet advanced in the afternoon upon the Newmarket Road from Richmond; made incessant efforts to break through the lines at this point; and the brigade double-quicked to support a battery, and formed, under the fire, as perfect a line as it would upon dress-parade. Gen. McCall's command [3/V], the Pennsylvania reserves, that had sustained the brunt of the attack, was hard pressed;[7] and the division [Hooker's] ran to its new position upon their left, and turned the current of the battle, which had commenced to flow against the Union forces. The regiment followed through the woods a narrow way which was thronged with the gunners and drivers of the "Dutch battery," who left their pieces upon the field,[8] and squads of infantry that

---

6 Willis Methodist Church, also referred to as St. Paul's.

7 Hooker noted the effect of the attack on the Pennsylvanians: "the enemy's attack had grown in force and violence, and after an ineffectual effort to resist it, the whole of McCall's division was completely routed, and many of the fugitives rushed down the road on which my right was resting, while others took to the cleared fields and broke through my lines from one end to the other." [Ibid. 111]

8 20-pounder Parrotts of Batteries A and C, 1st New York, attached to V corps.

were flying from the front. These circumstances did not dishearten the men who were marching by the left flank; and some who belonged to the companies upon the right rushed from their places to come in contact with the foe before their comrades. The colonel of one of the regiments that had been fighting[9] dashed to the rear upon his horse, before his command had been driven from its post, and excitedly screamed, "My men are all cut to pieces!" "Hurry, oh! Hurry, and save my poor men!" "There is one of them now, and wounded, too!" and seemed to be demoralized by fear. The troops double-quicked by him amid general laughter; and I heard a score of tongues utter remarks like these: "Dry up, you old fool!" — "Tear that eagle off your shoulders!" — "You ain't fit to be a private, you coward!" A number of swords that had been thrown aside by officers were scattered upon the ground; and, although I had recently exchanged my sash of worsted for one of silk,[10] the quickness of the movement did not allow me time to equip myself. A company of cavalry was deployed in the rear, and the commander trembled so much that he could not aim his revolver; and some vauntingly said, "Hooker's division don't need any cavalry to keep them in

---

9   The closest troops of McCall's (3/V) division to Grover's Brigade, holding the right of Hooker's battle line, was Seymour's Brigade (3/3/V) of Pennsylvania Reserves. [Sears 1988, 297]

10 Blake had been promoted to 2nd Lieutenant for gallantry. "I have been asked concerning my conduct in the battle of Williamsburg by persons, living remote from the strife, who evidently pictured me as acting like a hero of a dime novel. I had one answer only, that my duty was performed to the perfect satisfaction of my commanding officers, company and regiment." A committee in Dorchester eventually presented Blake with a sword, "the point of which was sharpened for desperate encounters, and although I cut the air, no human being was injured by its use." [Memoirs] The "dime novel" reference is to *The Young Lieutenant* (1865) by William Taylor Adams ("Oliver Optic"), the preface of which credits William Munroe and Henry Blake for valuable information on the movements of Hooker's division.

the front;" or "Our hands don't shake like that when we are there." An officer, carrying a saddle, came from the front, and remarked, in a tone of intense satisfaction, "I have done my share: I lost my horse, but I saved my saddle." These incidents, which may appear tame in their recital, amused the brigade which occupied with joyous cheers the position that had been assigned to it in the line.

The din of musketry and the cannonade, the yells of the rebels when they made a desperate assault, and the hurrahs of the Union soldiers when they were repulsed, did not cease until darkness covered the earth.[11] The *deep-throated engines*[12] upon the gunboats in the James River threw their monster shells into the ranks of the enemy, upon the left of the line, at five, P.M. The regiment held a road which the foe had entered in the afternoon; and many who had been lost in the confusion of the battle wandered about in the adjoining swamp,[13] from which a stream of prisoners was continually flowing into the ranks. Hundreds were yelling, in the peculiar effeminate voice of the Southerner, the names and numbers of their command, — "Third Alabama," "Seventh Georgia," "Sixth South-Carolina." A few soldiers stationed themselves in the advance, and sometimes shouted, "Here, by this oak;" "This way;" and captured a squad, who denounced the artifice as a "mean Yankee trick." The regiment

---

11 The newly-arrived 16th Massachusetts took the brunt of an attack by Pickett's Brigade, Longstreet's Division. "[They] received and repulsed the heaviest and most persistent attempts of the enemy to break the lines. … The Eleventh Massachusetts was thrown upon the extreme left of our division lines, in anticipation of an attempt to turn our flank [and] did not become engaged during the day." 1st and 16th Massachusetts took most of the 195 casualties in Grover's Brigade. [Grover, OR XI/2, 123]

12 *Paradise Lost*, Book 6 (Milton).

13 Theophilus Holmes' inexperienced Confederate division, recently arrived from the Department of North Carolina, was ineffectual against the Union left held by Porter's division.

took thirty prisoners, most of whom were delirious from the effects of whiskey, — wholly unable to point out a friend or foe, — and boasted that they had shot "heaps of Yankees." This startling fact explains the nature of that foolhardiness with which they charged upon batteries during the engagement. The rebel clothing (it could not be properly termed a uniform; for I did not see two persons that were dressed alike) was always faded to such an extent, that some skirmishers who wore shoddy, and necessarily shabby caps, were mistaken for the enemy, and fired upon by the men in the rear.

The troops marched towards Malvern Hill before daybreak [July 1st], without benefit of sleep; and the pioneers, who had partially cut the trunks of the trees which grew upon the sides of the road, waited for the column to pass them before they applied their final strokes. The army concentrated at Malvern Hill upon what the Union forces always seek, and the rebels avoid, — an open field. The appearance of the divisions, as they marched through acres of wheat which was ready for the harvest, and was garnered into the haversacks of famishing men, was inspiring to soldiers who had been placed, for a long period, in woods and swamps in which they could not see the right and left of the regiment. The bands, that had been dumb during the siege [of Yorktown], uttered the notes of patriotism, and revived the despondent; and cheers issued from the throats of thousands who deployed upon the plain, which was two miles in length and one in width, and supported three hundred cannon that defied the enemy. The signal-flags were disclosing the movements of the foe, and conveying orders from the roofs of houses upon the right and the decks of gunboats that protected the left; while the infantry, posted upon the commanding heights, had the confidence of [Confederate General] Stuart, who remarked to a [Union] prisoner, "If I had that hill, no army could drive me from it." The division was assigned to a position upon the left centre; and the hostile batteries debouched from a road at the distance of a mile, and concentrated their fire, a few minutes after nine, A.M., on July 1, upon the brigade when it was marching to this point. Some soldiers had taken the honey from seven beehives near a house: swarms of the exasperated insects stung the horses in the vicinity, with such serious results, that a battery, which had fought with valor the enemies of the country, was compelled to change its post; and mounted general and staff officers vigorously used their spurs to escape. The history of the day may be briefly described as a succession of desperate and reckless onsets upon various parts of the line, in which Lee was always unsuccessful: and his legions were slaughtered by the artillery,[14] including the siege-guns and those upon the monitors; while the Union loss was small, because the infantry was rarely engaged in close action. The incessant firing heated and discolored the pieces; and some rifled ordnance was rendered useless for accuracy, as it was constantly double-shotted. Quietness sometimes ruled during an hour, and no bullets would be discharged; but this was succeeded by the reverberations of cannon, which shook the earth in the concussion, although many who were not fighting, conquered by fatigue, slept upon their muskets, undisturbed.[15]

---

14 On the morning of July 1 Lee met with Jackson, Longstreet, Magruder, A. P. Hill, and D. H. Hill. The latter argued against an attack: "If General McClellan is there in strength, we had better let him alone." But Longstreet was unconcerned: "Don't get so scared, now that we've got him whipped." [B&L/2, 383] Despite the dominant Union position the Confederates made a series of mid-afternoon frontal attacks which were defeated with enormous losses. This outcome prompted D. H. Hill to bitterly remark that "it was not war; it was murder."

15 III corps was not involved in combat that day.

Upon July 2, drenched by the storm that always ensues after a great battle in which the forces of Nature have been violently discomposed, the army crowded in confusion upon a single road; and there was a moving mass of cattle, horses, and wagons, besides the infantry and batteries which belonged to different commands. Many excited disputes took place regarding the right to march in advance of the respective bodies of troops. The flying artillery of the [rebel] cavalry[16] threw a few shells into the bivouac of the brigade at Harrison's Landing; but this force was dispersed. The lines were established two days afterwards, and rifle-pits and redoubts were constructed during the succeeding month.[17] The official statements that Lee [had] commanded 180 or 200,000 men, while Gen. McClellan had only 75,000; the failure of the former to capture the extensive trains of wagons that filled every road, or penetrate the lines in a single instance, after suffering enormous losses, — inspired confidence in the general, who had won the glory of saving the whole force from destruction. There was also a feeling of disappointment at the result of the campaign, and grief for the fate of the sick and wounded who had been abandoned during the retreat. More than one-half of the [Union] prisoners that were taken in this movement after the battle of Gaines Mill [had] deserted from their companies, concealed themselves in the woods, and gladly yielded to the rebel cavalry; while others, who threw themselves upon the ground [during the Seven Days], and declared that they could not walk an inch farther on account of exhaustion, marched seventeen miles to Richmond within the succeeding twelve

---

16 J. E. B. Stuart's horse artillery occupied the heights overlooking the Union position at Harrison's Landing, firing every last shell it possessed over six hours. [OR XI/2, 519-520]

17 The Union position at Harrison's Landing was unassailable: from Kimage's Creek on the left to Herring Creek on the right, its flanks also protected by Union gunboats.

hours, with such rapidity that some of the guard fell out of the ranks. When the company arrived at Harrison's Landing, two men, who had only two pieces of tent, went to the forest to obtain shelter from the storm, and occupied the ground which had been selected by a brigadier-general for his headquarters.

"What are you doing here?" he asked, when he noticed them.

"We were going to fasten this canvas to the boughs, but didn't know that you were here," one of the privates replied, as they started to walk away. "You can stay here: this is my place; but I can move to the right," he said; and ordered the pioneers to pitch his tent in another spot. This gallant officer,[18] who recognized soldiers as human beings, displayed a kindness that was seldom exhibited by his peers.

Four soldiers, who reached Harrison's Landing before the regiment, crossed the [James] river in a boat, and were fired upon by some farmers, who held them as prisoners until a squad of [rebel] cavalry placed them upon their saddles behind themselves, and rode through the rebel encampments, in which the empty tents were standing to keep up appearances, while a guard of disabled men protected the property.

An excessive heat pervaded the camp;[19] but thousands had the privilege of bathing in the James, and enjoying habits of cleanliness, which the experience of Yorktown and Fair Oaks had taught them to value. Details were daily furnished to collect and burn the clothing which was cast aside on account of the vermin. Many officers tendered their resignations, which were generally

---

18 Most likely a reference to General Grover, whose treatment of volunteer soldiers was atypical of West Point officers.

19 For the rest of July the Army of the Potomac did nothing; its 100,000 men confined to an area about six square miles. In the confined space and summer heat, the health of the troops deteriorated.

disapproved; others feigned sickness to escape from the service; and one captain bribed two persons to carry him on a stretcher to the hospital-boat, and was absent from his regiment for more than a year. In addition to the list of ordinary diseases, soldiers died of the scurvy; and anti-scorbutic rations were issued to check this complaint.[20] The only event that disturbed the quietness of the camp occurred at midnight upon August 1, when the foe planted a battery on the south side of the James River, and opened upon the shipping and camps; but their guns were silent in half an hour.

The division and a brigade of cavalry moved toward Malvern Hill in the night of August 2;[21] but the guides misled the troops, and they returned to their quarters at sunrise. Gen. Hooker commanded the expedition, and resumed the march on the 4th, when the line of battle rested at midnight within a few rods of the hostile pickets. The force was in motion at daybreak; entered the road that passed by the church at Glendale; and attacked the rebels in their rear, at 5½, A.M. The [rebel] artillery opened upon the brigade with spherical case-shot; obtained an excellent range; one shell killing two,[22] cutting off the arm of another, and wounding four men, in one company of the

---

20 These rations had become necessary earlier in the Peninsula campaign. "They have just begun issuing that conglomerated mass called 'Dessicated Vegetables,' (the boys call them 'decayed vegetables'). They consist of carrots, turnips, potatoes, cabbages, onions, squashes, etc., etc., sliced up fine, mixed with glue(!) pressed into (awful) hard cakes about 20 by 30 inches, and then varnished! After eating a few meals, you begin to talk broken English (which is as far as I have got): I suppose, in time, we can talk Dutch fluently." *Chelsea Telegraph and Pioneer*, June 7, 1862.

21 To forestall orders from Lincoln to evacuate the Peninsula, McClellan launched an expedition to retake Malvern Hill. Hooker took his own division (2/III) and a cavalry brigade under Col. William Averell. [OR XI/2, 951] See Appendix: "Second Malvern Hill."

22 The two men killed on August 5 were Pvt. Edward F. Jones and Pvt. John Nolan of Company G. [MSSM/1, 792-793]

regiment: the troops pushed forward in four ranks, and sometimes dodged the balls; but none quit their places. A thick mist hindered a prompt advance; and, when the enemy was overpowered, only one hundred were captured, while the remainder, with the battery, retreated upon the James River Road. It was assumed that they could not escape, because a brigade had been ordered to seize this highway, and intercept them before the main body approached; but the plan failed in its execution through the base conduct of its commander, who maintained his reputation as a notorious drunkard. Gen. Hooker placed the officer in arrest, and remarked, in speaking of this action, "More prisoners would have been taken if that general had not been drunk."[23] He was the son of a well-known traitor in Philadelphia,[24] and received no punishment; and remained in the army until he committed suicide in a fit of *delirium tremens*. His death, in the language of the newspapers, was produced by an intense devotion to the service, and exposure in the performance of his duties.[25]

A trembling negro, who was paralyzed by fear, was shielded by a large oak, through the branches of which the shells were flying; and his frantic appeals for aid excited laughter, not grief, in the spectators who filed by him. The prisoners, like all that I saw, were extremely ignorant; not one in twenty being able to read and write: and their stolid faces showed a lack of mental

23 Brig. Gen. Francis E. Patterson of the Jersey Brigade (3/2/III). "This general — may as well be truly told — was on this occasion at least, too drunk to attend to his duties." [Haynes, 90]

24 Gen. Robert Patterson failed to keep Johnston's Army of the Shenandoah occupied in July 1861, thus allowing it to participate in the battle of Bull Run.

25 In November 1862, Francis Patterson, while under charges brought by Sickles for withdrawing his troops from Warrenton without orders, was found dead in his tent "from the apparently accidental discharge of his own firearm." [Sifakis, 304]

capacity which placed them upon a level with the natives of New Zealand. They were poorly supplied: some had pieces of carpet, which they used for blankets; and their bread was composed of flour mixed with water, which was baked upon a stick or the point of a bayonet. A woman, who lived near the picket-line, said that the rebels filled her house and begged for food after the battle of Malvern Hill; and they were so apprehensive of an advance, that Longstreet and Jackson prepared for action when the salute was fired in honor of the President [July 4th] at Harrison's Landing. The owner of the house, which had been the headquarters of Lee,[26] had posted up a notice that he did not wish to have any Yankees buried upon his land; and some soldiers who perused it applied the torch, and the splendid edifice, with its outbuildings, was completely destroyed. A squad of stragglers, who rarely render any service, made a charge with their unloaded muskets, captured seven cavalry-men, and rode upon their horses into camp, while the recent losers walked.

One of the prisoners [on burial detail], in answer to an inquiry about the grave of Major Chandler,[27] pointed out the spot in which a field officer had been buried; and the pioneers disinterred the body of a lieutenant-colonel, and found upon his person one hundred dollars: a strange fact, which amazed all who knew that the army thieves seldom missed one of the slain.[28] The position was evacuated upon the 7th [of August], and the old camp was again occupied. The exchanged prisoners rejoined

---

26 Lee was on the Willis Church Road during the Battle of Malvern Hill. His headquarters were not associated a specific building. There was a parsonage nearby, but it was used as a hospital and survived destruction.

27 Maj. Charles P. Chandler, 1st Massachusetts, was killed June 30 at Nelson's Farm, about two miles north of Malvern Hill.

28 Of the four Union lieutenant colonels killed during the Seven Days, only Eli T. Connor (81st Pennsylvania) died at Malvern Hill (the other three died at Gaines Mill). [OR XI/2, 985-992]

their commands from Belle Island; and their emaciated frames, and tales of suffering, had a good influence upon those who were inclined to prefer captivity to the chances of battle. Many of them stated, that, when the officer announced that a certain squad would be paroled on the next day, one hundred and twenty-one men saw the happiest moment of their lives: the [rebel] sergeant who had charge of it accepted bribes during the night, until the number was increased to one hundred and ninety-eight, and his haversack contained all the watches and valuables that the crowd possessed.[29]

The stores and the sick were sent upon transports to Fortress Monroe: the corps marched to Harrison's Landing on the 15th [of August], and proceeded *viâ* Williamsburg to Yorktown.[30] The [local] people openly expressed their joy at the failure of the retreating forces to capture Richmond: no guards were posted over rebel estates during the movement; the soldiers of Gen. Heintzelman's [III] corps made camp-fires of the well-seasoned fence-rails, and roasted the corn and potatoes which they took from the fields, without offending any of their generals.[31] They embarked on the 21st, and sailed up the Potomac to Washington in the crowded transports; and gun-barrels and bayonets glistened in every part of the vessel; and the bowsprit, shrouds, and rigging had a picturesque appearance.

---

29 3,021 Federals and 3,000 Confederates were exchanged August 3 at Aiken's Landing on the James River. [Denney, 73] Blake's story of bribery may account for the disparity in numbers.

30 After a month of inactivity, the Army of the Potomac was withdrawn from the Peninsula and ordered back to Washington to cooperate with the Army of Virginia.

31 So much for any concern about protecting Confederate private property. Maj. Gen. John Pope, commanding the Army of Virginia, issued General Order No. 5, directing his army to subsist on the country and issue vouchers which would be redeemed after the war to loyal citizens. [Hennessy 1993, 14]

Chapter VII

Bristoe Station,[1] Second Bull Run, and Chantilly

THE BRILLIANT REPUTATION which Gen. Pope acquired in the West,[2] and the energetic orders which he issued[3] upon assuming command in Virginia, delighted the armies upon the James and Potomac; and the highest confidence was placed in his military abilities. The division was packed into cattle-cars, inside and outside, on the steps and platforms; and a locomotive with the name and strength of "Samson" drew the regiment after sunset, upon August 25, upon the Orange and Alexandria Railroad, and arrived at Warrenton Junction at midnight. A regular camp was laid out on the 26th, near this point; shelter-tents were pitched; the blankets were spread upon straw; and the men retired to rest with the pleasant thought that they should soon recuperate, and purge the system of the scores and blotches with which the veterans of the Peninsula were afflicted. This agreeable dream was broken at four, A.M., on the 27th, by the voice of the orderly, which was always an unwelcome sound at this early hour. "Captain, captain! take three days' rations, fall in your company, and hold your men ready to start at a second's notice." A few reports of cannon were heard in the rear; and the division commenced to march upon the line of the railroad to

---

1　Also called the Battle of Kettle Run, which was about two miles southwest of Bristoe Station.

2　Pope's reputation was based on his three Western theater successes: the Battle of Blackwater Creek, the surrender of Island No. 10 in the Mississippi River, and the siege of Corinth.

3　Pope's initial statement to the Army of Virginia: "I have come to you from the West, where we have always seen the backs of our enemies; from an army whose business it has been to seek the adversary and to beat him when he was found; whose policy has been attack and not defense."

search for the ubiquitous Jackson, who had made a *détour* round the right of Gen. Pope, while that officer in his bulletins was driving him across the Rapidan with his cavalry and light artillery.[4] He [Jackson] burned the bridges over two runs, and secured trains that extended a mile upon the track, and were loaded with army-supplies of immense value. The force of Lee advanced in the mean while, upon the slopes of the Blue Ridge, to re-enforce this detachment; and the ingenious plan of operations, if executed with success, would have formed what was termed in popular language a "bag," which would have enclosed the main portion of the corps of Generals McDowell, Sigel, and Banks.[5] While certain officers with characteristic treachery[6] failed to move promptly from Alexandria, Generals Hooker and Kearney, whose loyalty was as conspicuous as their courage, pushed forward to the front; and the sanguine hopes of the rebels vanished, when they unexpectedly confronted the troops from the Army of the Potomac.

The division continued its march;[7] and the skirmishers advanced mile after mile beyond Catlett's Station without opposition, until an aide in the top of a tree reported that the

---

4   Union high command thought Jackson was still near the Rappahannock, or heading for the Shenandoah Valley. So it was a complete shock when his command of 15,000 turned up behind Pope's line. Jackson covered over fifty miles in two days, passing through Thoroughfare Gap in the Bull Run Mountains and cutting the Orange and Alexandria Railroad.

5   Corps commanders of the Army of Virginia: Franz Sigel, Nathaniel Banks, and Irvin McDowell.

6   Porter and Franklin were accused of foot-dragging because they felt an injustice had been done to McClellan in removing the Army of the Potomac from the Peninsula. Of course McClellan himself must be considered the chief foot-dragger.

7   Hooker's division had arrived by train at Warrenton Junction, but had to double back on foot toward Manassas Junction, which was the target of Jackson's flank march.

vedettes of the enemy were visible in the woods, and the brigade marched in line of battle through an orchard and a field of corn. The heat of the day and speed of the movement caused a perspiration which saturated the clothes as completely as the rain. The column was passing by a burning bridge about two, P.M., and the opinion was generally expressed that it was another raid, and the rebels were not within ten miles of their pursuers; but a shell burst at that moment over the heads of the debaters, and finished the discussion.[8] The Excelsior and Jersey brigades suffered severely in the fierce conflict which ensued; and the enemy was driven from the short pines in which the lines were concealed, after fighting an hour, and fled over the broad plain near Bristoe Station, while the brigade followed.[9] The cavalry afforded no assistance, because the commanding officer said the horses had no strength; and the infantry quickly marched by it. A battery which belonged to one of Pope's corps was as slow as the cavalry, and the captain of it acted like a person who did not wish to engage the retreating soldiers; and the men who had often witnessed with pride the prompt action of the artillery that formed a part of the division [Hooker's] viewed it with contempt, and scoffed at the members.[10] The hostile gunners entertained the same opinion, and did not notice the slothful battery, but directed their shell and shot at the advancing brigade, until they were compelled to withdraw to another position in the rear. The skirmishers fired at every suspicious-looking stump or bush

---

8   Jackson had moved the bulk of his forces toward Manassas Junction, leaving Ewell's division at Kettle Run, two miles west of Bristoe Station.

9   At The Battle of Kettle Run, Hooker sent three of the Jersey Brigade's regiments straight up the railroad, while the Excelsior Brigade advanced on the left. Grover's Brigade was in reserve, with only the 2nd New Hampshire sent in to support the left toward the end of the battle. [OR XII/2, 444]

10 The artillery of Hooker's division had not fully arrived from Alexandria.

when they ascended the rising ground; and the most anxious moments of the day were those which their cautious steps occupied in approaching the crest which might shelter the enemy.

The sudden onset by Gen. Hooker had not been foreseen by Jackson; and the appearance of the field showed that it was a hasty fight. The dead and wounded had been abandoned; knapsacks and equipments were scattered upon the plain; beeves had been killed; the fires were burning beneath the Dutch-ovens, which contained baking bread or roasting meat; dough was left in the pans, and dinners had been prepared in the houses for the officers. There were also two bags of raw peanuts, from which the rebel cooks manufactured a substitute for coffee. The civilians, who never gave or sold food to Union soldiers, had collected geese, turkeys, and the "fatted calf," for their friends: and one woman cooked two barrels of cakes "for family use," so she said; but they were devoured by the victors. The horses for the [Union] field and staff officers had not been transported from Yorktown, so that they were compelled to march on foot; and although they always declared that it was more fatiguing to ride than walk, and mercilessly shouted to weary men, "Close up," or "Double-quick," they were the first persons who left the ranks: and most of the regiments were commanded by captains during the engagement. The number of stragglers was very large,— one-fourth of the effective strength of the force; and squads of skulkers, who were utterly exhausted by the heat whenever a bullet whistled near them so that they could not creep to the front, ran to the rear, placed [percussion] caps upon the nipples of their muskets, blackened their faces and mouths with powder to resemble those who had been engaged, and rejoined their companies after the battle, and explained matters by saying that they fought in another portion of the field.

Among others at the hospital was a German, who was mortally wounded, and said that there were two companies of his countrymen in the rebel regiment who had been forced to leave their workshops, and enlist in the army. He gasped, in his broken language, "Oh! how hard to die, when I have been in this land only three months!" The prisoners were constantly talking about the good qualities of their commander, who had marched them sixty miles in the last two days; would seize Washington within a week; and one of them exclaimed, "If your generals were as smart as Jackson, you would soon conquer us." A house inhabited by an Irish family was exposed to the shells during this contest; and the wife entered the closet, and prayed to the Virgin for safety, while her husband and children remained in the cellar. The soldiers took all the clothing they could discover in one building, from which the general had been fired upon; and the owner remarked to one of them,—

"You have got on my shirt."

"Yes; and I intend to keep it on."

They also ransacked another house, in which some Federal uniforms, stained with blood, were found; and a light-fingered man stole the spectacles from the nose of an aged citizen, who pretended to know nothing respecting them, and complained about the treatment which he had received. Hundreds of baggage-cars, with their valuable contents, were burning five miles in our front [at Manassas Junction]; and the skies were darkened in the afternoon by a dense cloud of smoke, which was a flame of fire at night that lighted up the heavens.

Jackson retreated from[11] Manassas Junction during the following day, and subsequently formed his line of battle upon the old field of Bull Run. Gen. McDowell [Third Corps, Army of

---

11 Original text says "to."

Virginia] was ordered to hold Thoroughfare Gap, a position of great natural strength in the Bull-Run mountains; but Longstreet, in the evening, with a few puffs of smoke from the rifles of his sharpshooters, easily gained possession of this outlet; and Lee was allowed to re-enforce Jackson with his whole army, without the slightest opposition. A portion of Gen. Pope's troops marched, on the 28th, by the division, which followed them along the line of the railroad, halted at night, moved from the bivouac at two, A.M., of the 29th, and stood upon the heights of Centreville an hour after sunrise. Matches were very scarce upon this campaign; and a private who intended to light one gave public notice to the crowd, who surrounded him with slips of paper, and pipes in their hands. Some soldiers were in a destitute condition, and suffered from blistered feet, as they had no shoes; and others required a pair of pants or a blouse: but all gladly pursued Jackson; and his capture was considered a certain event.[12] The column cheered Gen. Pope when he rode along, accompanied by a vast body-guard, and responded, "I am glad to see you in such good spirits to-day." Justice obliges me to write, that, after the experience of one of his mismanaged battles, the silence of deep contempt was the sole greeting that he perceived in the faces of these disappointed soldiers. Three miles from the battle-field, the division met a squad of five hundred cowards, who had been paroled because the enemy could not, or would not, issue rations to them;[13] and they exultingly boasted that their

---

12 Pope believed he had Jackson completely surrounded and declared that, with prompt action "We shall bag the whole crowd."

13 The Union parolees were probably the defenders of Manassas Junction, some 115 men of the 105th Pennsylvania, a detachment of cavalry, and eight artillery pieces. They had been overrun by Trimble's brigade the night of August 26-27. The balance of the prisoners would have been from the Taylor's brigade (1/1/VI), who had attempted on August 27 to retake Manassas Junction from the east, and had been mauled by the

lives were safe, as they would not be compelled to go into the pending conflict. Most of this number had been detailed to guard the railroad or the trains, and basely[14] surrendered their posts without firing a musket to alarm their companions or check the foe. Many of the dastards exchanged suits with the rebels, and wore the butternut clothing; while the latter arrayed their spies in the Federal uniform, and gained other advantages by using them.

The stream [Bull Run] was forded [August 29th], and the graves and bones of the dead, rusty fragments of iron, and the weather-beaten *débris* of that contest, reminded the men that they were again in the midst of the familiar scenes of the first battle of Bull Run. The cannonading was brisk at intervals during the day. Large tracts of the field were black and smoking from the effect of the burning grass which the shells ignited; and a small force was occasionally engaged upon the right: but there was no general conflict.[15] The brigade took the position assigned to it, upon the slope of a hill, to support a battery which was attached to Sigel's corps; and no infantry was visible in any direction, although the land was open, and objects within the distance of half a mile were readily seen. There was no firing, with the exception of the time when the troops debouched from the road in the morning; and the soldiers rested for hours until four, P.M. At this moment, the enemy opened with solid shot upon the battery, which did not discharge one piece in response; the drivers mounted their horses; all rushed pell-mell through the ranks of the fearless and enraged support, and did not halt within

---

Confederates. [Hennessy 1993, 114, 124-127]

14 Original text says "barely."

15 Pope's battle plan for the 29th called for the corps of Porter and McDowell to attack the right of Jackson's line, while Sigel's held Jackson in place with demonstrations in front. Pope was unaware that Longstreet had arrived alongside Jackson, blocking this move. In the early afternoon, Grover's Brigade was temporarily placed under the direction of Sigel.

the range of the artillery from which they had so cowardly fled. A member of the staff, dressed like an officer of the day, immediately arrived, and gave a verbal order to the brigade commander [Grover]; after which the regiments were formed and marched, unmindful of the cannon-balls, towards the right of the line, and halted in the border of a thick forest, in which many skirmishes had taken place.

"What does the general want me to do now?" Gen. Grover asked the aide who again rode up to the brigade.

"Go into the woods and charge," was the answer.[16]

"Where is my support?" the commander wisely inquired; for there were no troops near the position.

"It is coming."

After waiting fifteen minutes for this body to appear, the officer returned and said that "the general was much displeased" because the charge had not been made; and the order was at once issued, "Fix bayonet." Each man was inspired by these magical words; great enthusiasm arose when this command was "passed" from company to company; and the soldiers, led by their brave general, advanced upon a hidden foe, through tangled woods which constantly interfered with the formation of the ranks.

"Colonel, do you know what we are going to charge on?" a private inquired.

"Yes: a good dinner."

The rebel skirmishers were driven in upon their reserve behind the bank of an unfinished railroad; and detachments from five brigades were massed in three lines under the command of Ewell[17] to resist the onset of the inferior force that menaced

---

16 Hooker agreed to the attack by Grover's Brigade only after he had convinced Pope that Kearny's division (1/III) should cooperate on the Union right. [Hennessy 1993, 245]

17 Grover's Brigade was attacking the troops of A. P. Hill, not Ewell.

them. "We will stir up these fellows with a long pole in a minute," one of the company said when the bullets began to sing; and he welcomed the fatal shot which cut him down in his youth. "Victory or death" were the last words of another humble hero. The awful volleys did not impede the storming party that pressed on over the bodies of the dead and dying; while the thousands of bullets which flew through the air seemed to create a breeze that made the leaves upon the trees rustle, and a shower of small boughs and twigs fell upon the ground. The balls penetrated the barrels and shattered the stocks of many muskets; but the soldiers who carried them picked up those that had been dropped upon the ground by helpless comrades, and allowed no slight accident of this character to interrupt them in the noble work. The railroad bank was gained, and the column with cheers passed over it, and advanced over the groups of the slain and mangled rebels who had rolled down the declivity when they lost their strength. The second line was broken; both were scattered through the woods; and victory appeared to be certain, until the last support, that had rested upon their breasts on the ground, suddenly rose up and delivered a destructive volley, which forced the brigade, that had already lost more than one third of its number in killed and wounded, to retreat. Ewell, suffering from his shattered knee,[18] was borne to the rear in a blanket, and his leg was amputated. The horse of Gen. Grover was shot upon the railroad bank while he was encouraging the men to go forward; and he had barely time to dismount before the animal, mad with pain, dashed into the ranks of the enemy. The woods always concealed the movements of the troops; and at one point a portion of the foe fell back, while others remained. The forces sometimes met face to face, and the bayonet and sword — weapons that do not pierce

---

18 Ewell's wounding actually occurred the day before.

soldiers in nine-tenths of the battles that are fought — were used with deadly effect in several instances. A corporal exclaimed in the din of this combat, "Dish ish no place for de mens," and fled to the rear with the speed of the mythical "flying Dutchman."[19] In one company of the regiment, a son was killed by the side of his father, who continued to perform his duty with the firmness of a stoic, and remarked to his amazed comrades, in a tone which showed how a strong patriotic ardor can triumph over the deepest emotion of affection, "I had rather see him shot dead as he was, than see him run away."[20]

The victors rallied the fugitives after this repulse, and their superior force enabled them to assault in front and upon both flanks the line which had been contracted by the severe losses in the charge; and the brigade fell back to the first position under a fire of grape and canister which was added to the musketry. The regimental flag was torn from the staff by unfriendly limbs in passing through the forest, and the eagle that surmounted it was cut off in the contest. The commander of the color-company saved these precious emblems, and earnestly shouted when the lines were re-formed, "Eleventh, rally round the pole!" which was then, if possible, more honored than when it was bedecked in folds of bunting. Gen. Grover, who displayed the gallantry throughout this action that he had exhibited upon the Peninsula, waved his hat upon the point of his sword to animate his brigade and prepare for a renewal of the fight. Many were scarcely able

---

19 Any corporal Blake overheard in the din of combat would have been one in his own company. Blake gives him a German accent, so he may be referring to Cereah Schnepf, who survived the battle but deserted the following year, during the march to Gettysburg. [MSSM/1, 813]

20 The father and son were Rodney and Con Edmands (ages 44 and 18) from Reading. Rodney kept going until discharged for wounds September 3, 1863. He quickly recovered and enlisted on December 30, 1863 in the 59th Massachusetts Veteran Regiment. [MSSM/1, 803]

to speak on account of hoarseness caused by intense cheering, and some officers blistered the palms of their hands by waving swords when they charged with their commands. The support was not present when the soldiers emerged from the woods, although an hour had elapsed since the aide stated that it was "coming." Another brigade soon reached the scene, and made a charge over the ground which had been recently won and lost;[21] but was repulsed before the railroad bank was attained. The motives that governed the officer in command who caused this large destruction of life were never understood by the fortunate survivors, who agreed with Gen. Hooker when he protested against the proposed movement as "a useless slaughter of my men to attempt to win a position which was of no military value when it was gained."[22]

The enemy followed the retreating troops after the disaster; and the brigade retired so that the next contest would occur in the open field: but the rebels, who did not wish to leave their shelter, halted in the fringe of the forest, and formed an excellent line, while the "stars and bars" that glittered upon their brilliant crimson flags resembled the vivid hues of the most venomous serpents. The commander of the mountain howitzers[23] promptly obeyed the order to "pring up de shack-asses;" the impatient cannoneers stood by their pieces, and urged the soldiers who

---

21 Nagle's Brigade (1/2/IX).

22 Grover's Brigade lost fully one third of about 1,500 men. In the 11th Massachusetts, Blaisdell reported that "out of 283 officers and men who participated in the fight, 3 officers and 7 enlisted men were killed, 3 officers and 74 enlisted men were wounded, and 25 missing ... all in the space of fifteen or twenty minutes." [OR XII/3, 441] See Appendix: "Grover at Groveton."

23 Blake gives the commander of the howitzers a German accent, which suggests they may have belonged to a battery from Sigel's predominantly German corps. See Hennessy for some of the candidates. [Hennessy 1993, 257f]

were marching in front to hasten to the rear, so that they could open; and the warning, "Get out of the way, or we'll blow your head off!" developed a new energy in many weary muscles. The splendid front was broken by the rounds of canister, and quickly disappeared in the forest, and left a line of skirmishers, who shot all the wounded that attempted to crawl from their exposed positions upon the field to the Union pickets. The men slept with comfort at night upon straw which had been taken from the same stack that stood upon the ground in the action of July 21, 1861; and some, who knew that a bullet had penetrated their blankets or great coats, which were tightly rolled and fastened to the knapsacks, found that one hole became thirty or forty when they were spread out for use.

The pickets were unusually quiet: strong re-enforcements arrived in the morning for both armies, and all expected a glorious result; but I was soon convinced that no troops however large in number, could contend against Lee with success while Generals Pope and McDowell commanded them. If Gen. Fitz John Porter, who received a lenient punishment for the crime which he committed,[24] and other officers of high rank, who merited the same justice for acts equally culpable, had taken part in this battle, the same causes would have produced the inevitable defeat. Gen. McDowell was viewed as a traitor by a large majority of the officers and men, and was distrusted by officers upon his staff, by members of his body-guard, and those who were constantly associated with him; and thousands of soldiers firmly believed that their lives would be purposely wasted if they obeyed his orders in the time of conflict. From prisoners I ascertained that the rebel army entertained the same

---

24 Porter was accused by Pope of failing to obey orders on the 29th. As a devotee of McClellan, Porter's court-martial conviction in December 1862 was based more on political than on military considerations.

idea; and Lee knew that thirty thousand men of the force in his front were demoralized on this account: and the battle that followed proved that it was a melancholy fact. Gen. Pope acted like a dunderpate during the day (the 30th), and scorning the wise advice of abler generals, like Hooker and Kearney, allowed Gen. McDowell to maneuver the troops upon the field. I boldly declare that the task would have been discharged with greater ability by intelligent sergeants in the regiments; and the results were perfect illustrations of those which ensue "when the blind lead the blind."[25]

The hours quietly passed away, with the exception of an occasional firing by the skirmishers, until four, P.M.; and many batteries and brigades were marched to the left, to that plateau near the Henry House which was the scene of the heaviest fighting in the old engagement. The national forces were carelessly deployed upon the cleared land, so that Lee, from a commanding hill, could perceive and inspect the number and position of every Union regiment and battery; while he massed his divisions in the woods, and it was impossible to see any regular body of them. Thousands of the infantry rested behind their stacks; and some batteries were never unlimbered, and rendered no service, although they were often required to prevent the shameful defeat that followed. The enemy concentrated his strength upon his right, made a feint upon the centre with a small force, and suddenly overwhelmed the left, which was composed of Gen. McDowell's corps, brigades of which fled in confusion after receiving one volley, and did not attempt to re-form, but shouted defiantly to their commanding officers, "You can't play it on us!" and similar cries. The troops comprising the right wing, which were posted one mile from this point, stood upon fences

---

25 Biblical reference to Matthew 15:14

and the wheels of gun-carriages, and watched the struggle with the keenest interest until they were satisfied that the day was lost.

When the eye excluded the smoke and havoc of the conflict, and gazed upon the scenery,— the green belts of the forest, the undulations and heights upon the field, the cloudless skies, and the distant summits of the Bull Run and Blue Ridge that formed the back-ground of the view; and — *Blue against the bluer heavens Stood the mountain, calm and still*[26] — the soul was enchanted with the unsurpassed beauty of Nature. In the midst of this loveliness, the scenes of horror upon the plain — the mutilated forms of suffering men, the prolonged roll of musketry, the reverberations of the artillery, the yells of the rebels when they charged and captured a battery, and the sulphurous smoke that at times enveloped the combatants — presented a terrible contrast.

Cattle had been killed, and issued to the brigade, and many were broiling the beef over the fires while the contest was undecided upon the left. The exploding shells continually emitted globes of smoke; and the difference in the color showed that the enemy used the finest quality of powder, which was white, while the other was black. Pieces of railroad-iron, that rushed with an irregular motion through the air, indicated a limited supply of ammunition. Three hours vanished while the brigade was alternately double-quicking upon the field, or halting for a brief period to support the artillery, but steadily approaching the left, and fearing the canister that was hurled over it from the batteries in the rear more than that of the enemy in front. The regiment at one time held the same position upon the Leesburg Road which it had defended in the first action of Bull Run; and history narrates

____

26 *Listening Angels* by Adelaide Anne Procter (1825-1864), whose poetry was
    first published by Charles Dickens in 1853. Her best-known poem is *A
    Lost Chord*, which was set to music by Sir Arthur Sullivan in the 1870's.

few coincidences that are stranger than this. The field was abandoned while the sun was sinking beneath the horizon; and the column marched through the cold water of the runs, and bivouacked near Centreville at midnight [August 30th]. All were affected with grief by this disaster, and I noticed officers who restrained with difficulty tears of sorrow; and general indignation was expressed against the two commanders who were responsible for the useless effusion of such precious blood.

When the brigade had retreated a short distance, it passed by Gen. McDowell, who sat upon his horse in the road; and the most profane oaths were uttered in reference to his conduct, and his ears must have often caught the insulting taunts of thousands of brave and patriotic men. There was scarcely a moment during the march in which I did not hear the epithets "villain," "traitor," or "scoundrel," applied to his name. He wore a hat made in such a peculiar style, that he could be identified by the ranks of the contending armies. This strangely fashioned article was not a part of the Federal uniform; and while Gen. McDowell knew that he had no right to wear it, and would have roughly censured an officer, if he had noticed, upon inspection, any volunteer who was clothed in this outlandish apparel, the suspicions of those who doubted his loyalty were increased by this gross violation of military regulations. "I would sooner shoot McDowell than Jackson!" "How guilty he looked with that basket upon his head!" "It is an outrage to put men under that traitor to be murdered!" were the remarks which were constantly repeated. "My men went upon the field as if they were upon dress parade; but, in a few minutes, I was left all alone," Gen. McDowell said to Col. Marston, who commanded the Second New Hampshire Volunteers that marched in the rear of the regiment.

Incidents were hourly witnessed that will be remembered as long as the mind retains its faculties; and a record of some of them may interest the reader. Gen. Hooker astonished certain officers of high rank in Pope's army by displaying an example of courage which they should have followed; and one of them asked, "Who ish dat general mit a white horse and red face? He cares nothing for bullets." Untaught by the disastrous results of the battle of last year, batteries without an adequate support were pushed to the front in the same heedless manner, and upon the same ground on which those of Griffin and Ricketts were lost; and Lee captured them with ease. Many of the officers and cannoneers escaped, and their statements added fuel to the flame that was already consuming the reputation of the person I have so often named. "Mein Gott, mein Gott, general, the rebels will have mine every piece!" one artillery commander exclaimed; while another, wringing his hands with anguish, shouted many times, "All my guns lost, all my guns lost, through that infernal — McDowell!" — "Sergeant," said a gray-haired brigadier to a non-commissioned officer of the regiment, who was wounded, and traveling to the hospital, "how are things going?" — "We hold our own now; but McDowell has charge of the left," he replied. "Then God save the left, if McDowell has charge of it!" the general answered in a tone of utter despair. A general who belonged to the exceedingly small circle of McDowell's military admirers, deserted his men, rushed to one of the hospitals, and yelled, "Two hundred rebel cavalry are driving my brigade! Can't you help me?" His force consisted of three thousand infantry; and the wounded indignantly insulted him: "Go back to your command, you coward"! "Shoot the skedaddler!" and he rode still farther to the rear.

Scenes illustrating the extremes in human character occurred; and there were mean subterfuges to evade, and noble efforts to brave, the dangers of battle. An artillery officer was groaning, and seemed to suffer intense pain, until a shell burst near him; when he jumped from the stretcher, and fled so swiftly, that those who were carrying him could not keep pace with his flying feet. A captain in one regiment skulked out of the fight, and passed by the provost-guard by showing his hands, which were covered with blood that had flowed from the wounds of one of his company. If officers were shot, or relaxed their vigilance, squads of three or four soldiers would leave the ranks, and carry a disabled man or escort an unarmed prisoner to the rear. A corporal, scorning aid, used his musket as a crutch, and walked to the hospital; and one of the company, who was mortally wounded, implored his comrades, who had taken him from the ground while the brigade was retreating, to escape, as they might be captured by the enemy, and he did not wish them to suffer the privations of prisoners.[27] Shoes and articles of clothing were thrown away by some to enable them to shirk their duty; and others, who were actually destitute, dragged themselves upon swollen and blistered feet, to the front. Many soldiers were drunk in Alexandria while their comrades were dying upon the field; and the number that fought, if compared with the rolls of those who were paid, reveals a lax state of discipline. In one regiment, only 302 men in 843 were present during the action; in another, consisting of 847, only 318 took part; in two regiments the ratio was smaller,[28] and 596 were engaged: so that less than three-eighths of a brigade performed the hazardous duty of fighting. The rebels advanced their lines, threw shells into the hospitals, and killed soldiers who were helpless on account of wounds. The

---

27 Possibly Capt. Benjamin Stone, who was left on the battlefield for days. When the Confederates abandoned the field, Stone was picked up and sent to a Washington hospital, but he died of his wounds September 10. [HDS]

28 Blake is probably referring to the "present for duty" figures in Grover's Brigade. Blake does not exaggerate the disparity between the number enrolled and those actually present, which was a problem throughout the war.

frightened surgeons and [male] nurses abandoned, in one place, those who required their care; while the so-called daughters of two regiments boldly remained,[29] and loudly denounced the runaways. An orderly in the regiment found a scabbard which was besmeared with blood, and a private discovered the sword that was a slight distance from it; and, by pitching a copper, both were won by the sergeant, who was afterwards promoted, and wore them until he was killed at Chancellorsville.[30] The fragments of exploding shells could be easily discerned in the air, and I noticed one which shattered the jaw of a bugler as he was sounding a call. The sufferings of those who were captured by the enemy cannot be described: the wounded had no care during five days; and others were reduced so much by insufficient food, that, when they were released, they gladly ate the crumbs of hard bread which had been scattered more than a week upon the ground at Centreville. The corpses of three hundred soldiers were placed upon each other, and buried by throwing earth upon them; so that they were lightly covered. The number of ambulances was inadequate to convey the wounded that had been paroled; and two hundred hacks and carriages were seized in the streets of

---

29 Harriet Dame accompanied the 2nd New Hampshire in many of its campaigns. At Second Manassas she fell into Confederate hands, but was released because she assisted Union and Rebel soldiers alike. [Holland, 71] Another nurse, Helen Gilson, was mentioned by Chaplain Cudworth of the 1st Massachusetts: "She has been with the Division ever since the disastrous summer of '62, and has greatly endeared herself to officers and men. To probably a hundred or more of our gallant boys, on the battle-field and in the hospital, hers has been the last face their dimming eyes have seen, hers the last words poured into their dying ears. At all times thoughtful and attentive, her special worth has shown itself during and after our battles, or when fever has raged through our camps. To thousands, then, she has proved herself a ministering angel indeed." *Chelsea Telegraph and Pioneer* April 30, 1864

30 The only member of the 11th Massachusetts fitting this account is Sgt. John S. Harris. [MSSM/1, 783]

Washington, and their drivers were compelled to go to Manassas for this humane purpose. The rebel prisoners facetiously remarked, that, upon the campaign, "Jackson did all the praying, while Ewell did the swearing."

The army rested upon the heights of Centreville during two days, and enjoyed the comfort of the barracks which had been occupied by the force of Johnston[31] in the winter of 1861-2; and the enemy showed no inclination to storm the works which they had constructed for their own protection. Lee gained the position which bore the historic name of Chantilly.[32] Gen. Pope discovered that he was flanked, and ordered the divisions of Generals Kearney, Hooker, and Reno,[33] to march to this point. Before the movement was commenced, Gen. Kearney made a speech of exhortation to his men for the last time; and the troops, in the midst of the storm and darkness, advanced through the forests, and fields of corn, in which a few regiments suffered severely; but the brigade [of Grover] was posted upon the left, and its skirmishers were unmolested. Generals Stevens and Kearney were killed;[34] but the foe was speedily driven from the position, and the line of retreat between Washington and Alexandria was secured. One-half of the regiment was kept under arms during the night; while the remainder, trembling with cold, attempted to sleep in the rain and mud. The whole force was in motion at twilight, and encamped at Alexandria upon September 3.

---

31 Original text says "Johnson."

32 In the final phase of the campaign against Pope, Jackson attempted again to flank the Union army at Centreville. The battle of Chantilly ensued on September 1, near what is now Dulles Airport. Blake calls it a "historic name" as its namesake was the *Château de Chantilly* in France.

33 Kearny led 1/III, Hooker 2/III, and Reno 2/IX.

34 Generals Isaac Stevens (IX) and Philip Kearny (1/III).

The Government acted with decision, and justly deprived Generals Pope and McDowell of their powerful command; and the first was banished to the frontier of Minnesota, while the last was not entrusted with any military power until Gen. Grant exercised his usual sagacity, and exiled him to California.[35] These just measures pleased the unfortunate soldiers who had been compelled to obey their orders; and the appointment of Gen. Hooker to command the first corps, *vice* McDowell relieved, was received with joyous shouts and cheers;[36] and the wisdom of those that made this important change was vindicated by [his] brilliant conduct at South Mountain and Antietam.

The army mourned the national loss of Major-Gen. Kearney, who was killed at Chantilly; and his memory will be cherished as long as exalted patriotism, inspiring courage, and justice towards men, are revered by mankind. Qualified to be the head of the army, he accepted command of a brigade. Leaving the comforts which his large wealth afforded, he welcomed the most trying hardships of the service. In another zone, the enemies of his country had taken his arm;[37] but his zeal triumphed over the disability, and he fought until he had sacrificed his life. Placing the reins between his teeth, and grasping in his single hand the two-edge sword, he led his men in the charge that was never checked. Humane to those who were his inferiors, the orderlies were directed to bring water in canteens to the soldiers when the exigencies of the hour required that all should remain in the ranks at the front. Impetuous in thought and action as the flash of

---

35 Pope requested reassignment, and soon was sent to the Department of the Northwest. McDowell retired from active duty; in 1864 he was assigned to the command of the Department of the Pacific.

36 Pope's Army of Virginia was incorporated into the Army of the Potomac under McClellan.

37 At the Battle of Churubusco (Mexico), Kearny had been wounded in the left arm, resulting in amputation.

his fiery eye, he censured with the same vehemence the misconduct of a private, or the general of the highest rank in the Union forces. Beloved by his division, the red badge which he instituted[38] was always worn by the officers and men with the same proud feeling with which the heroic commander displayed the cross of the Legion of Honor, which never enrolled a nobler chevalier. Bravely performing his public tasks, the death of the pure patriot and consummate soldier was a fitting conclusion of his eventful life.[39]

---

38 The men of Kearny's division wore a red diamond-shaped cloth badge on their caps, the precursor to the familiar corps badges instituted under Hooker in 1863 for the entire Union army.

39 Kearny served in the French and United States armies. In 1859, serving with the *Chasseurs d'Afrique,* he was awarded the cross of the Legion of Honor for gallantry on the field of Solferino. At the battle of Chantilly, Kearny inadvertently entered the Confederate line and was trying to escape when he was shot through the spine and instantly killed. His remains were returned by Lee under a flag of truce. [Union Army, Vol. 8]

Chapter VIII

The Battle of Fredericksburg

THE DIVISION had marched from Warrenton Junction [on August 27th] with the expectation that it would soon return to the camp; and the guards who had charge of the munitions of war and private property destroyed them; and there was hardly an officer in the command that possessed any clothing, besides what was on his person, when it reached Alexandria.[1] The troops daily labored upon the earthworks; and the order that was issued by the President at this time, prohibiting work upon the Sabbath, was always disregarded.[2] The privates of the garrisons in the forts were dressed in better apparel than the officers of the old regiments, viewed with disdain their tattered appearance, and played cricket[3] and similar games for exercise, while the veterans from the Peninsula used the spade.[4] The engineers who resided in Washington rode around the works once in three days in an elegant carriage, and gave directions, according to their caprices, to the officer in charge of the fatigue-parties to cut down a few

---

1 Hooker had written that 2/III division was so demoralized by the severe losses they had sustained so far in 1862 that it was "in no condition to go into battle at this time." [OR XII/2, 437] III corps remained in Washington during the Antietam Campaign. Grover's Brigade was initially stationed at Fort Lyon in Alexandria, but soon moved to the vicinity of Fort Ward for the month of September.

2 Lincoln did not issued the Executive Order *Respecting the Observance of the Sabbath Day in the Army and Navy* until November, perhaps after witnessing the Sunday work details that September.

3 Cricket was quickly being overtaken by baseball in popularity. [Kirsch]

4 "The brigade built a line of rifle pits between Forts Ward and Worth, and pickets and fatigue duties combined became so excessive as to cause much dissatisfaction among the men; especially as much of the work in both directions was more a matter of furbelow than of utility." [Haynes, 141]

stumps, or remove an inch of gravel from the crest of a parapet, or increase or diminish the angle of one of the slopes five degrees. The brigade and division had lost their generals by promotion,[5] and picket-duty was performed under the supervision of a coward who never visited a post if the enemy was near it.[6] He inspected the lines in places of safety with great pomposity, and prohibited the use of lights at night, although those [national troops] who were thirty miles from Alexandria had blazing camp-fires.

Inspections, reviews, and similar forms of camp-life, constantly took place; and I was sometimes entertained by the comments of a commanding officer who examined a regiment, and seemed to be determined to find as much fault as possible. If the magazines had been supplied with ammunition, he said, "You coward! why didn't you fire some cartridges in the fight?" If they had not been procured, the men were reproved for their negligence. If a thin coating of dust was visible upon the head of a rammer, he remarked, "That gun is a solid bar of iron;" "The rust is six inches deep in the barrel;" or "You might as well try to shoot with a tree." The "good and holy man," a chaplain, held religious services upon September 21, in compliance with

---

5   Hooker was promoted to the command of I Corps, leading it at South Mountain and Antietam. Grover was given command of 4th Division, XIX Corps, Department of the Gulf. Maj. Gen. Daniel Sickles replaced Hooker in command of 2/III division.

6   The craven picket-duty supervisor was Joseph B. Carr, the new brigade commander. This is the first of Blake's many caustic references to Carr, but we are not told precisely why or when Blake developed his intense dislike. Carr's confirmation by the Senate was lengthy: first promoted to brigadier general 9/7/62, not confirmed; reappointed 3/30/63, not confirmed; reappointed 4/9/64, confirmed 6/30/64, back dated to 3/30/63. After Gettysburg, when he was acting division commander, he was in the odd situation of being junior to his subordinate brigade commanders.

orders;[7] and, although no congregation was present,[8] read his prayers in a loud voice, and seemed to be satisfied with his endeavors.

The camp was broken up on November 1, when the brigade bivouacked at Fairfax Court House; and the command halted upon the 3d at Manassas Junction, near which many barracks of logs, and chimneys and ovens of red sandstone, were standing. A few [rebel] guerrillas captured a wood-train the day before the regiment arrived; and a youth of thirteen years of age, who lived near this point and saw the affair, spoke with much frankness about the base conduct of the guard, which was composed of sixty men from a New York regiment in Gen. Sigel's corps; and his recital of the facts was confirmed. "They were a lot of cowards," he earnestly said; "and four of them hid in the culvert, and came out after our cavalry had gone, and told me and my brother (pointing to a boy of about ten years old) not to kill them, for they were our prisoners: and I told them to keep still, for the cavalry might hear them, and come back and get them. The others ran away, and kept up with the horses; and, if they had run as fast t'other way, the cavalry wouldn't have got any of them." The brigade bivouacked at Bristoe Station upon the 7th, in the first snow-storm of the season, and met a portion of the division[9]

---

7   Sunday religious services may have been ordered in order to recognize Union accomplishments of the Antietam campaign.

8   Blake's comment about Sunday work details might lead one to think that the men were out digging earthworks again. Or it may have been that they just didn't like Watson's preaching. And despite the fact that the the most notorious red-light district in Washington was known as "the hooker division," there is no indication in Blake, Haynes, or other sources that enlisted men were in Washington validating the unsavory name.

9   The Jersey Brigade under Brig. Gen. Francis E. Patterson (3/2/III). Sickles reported November 7th that Patterson had retreated without orders. Patterson, in reporting the arrival of enemy troops, said that without support "My position is untenable. The whistles of cars are going freely,

that was retreating from Warrenton Junction with the news, which their general had communicated, that the rebels had a large force at that place, and it was considered foolhardiness for a small body of troops to attack them. There was no opposition upon the succeeding day, when the column advanced,[10] and no signs of a recent occupation by the enemy could be perceived. The troops had not halted an hour before the citizens in the vicinity visited the camps to purchase salt, and other articles of food. Contrabands, carrying small packs of clothing, were continually passing over the railroad to Alexandria during the period in which the brigade held this post. The train that conveyed Gen. Hooker to Warrenton, where the army had been concentrated, stopped at the Junction upon the 11th, and the troops received him with loud cheers. Gen. Halleck, than whom no officer was more universally detested by the soldiers, arrived upon the following day; but not one voice of welcome was heard in the large number that knew he was present. Three hundred rebel prisoners were transported upon the railroad on the 13th; and interesting conversations ensued, in which opinions were expressed about different generals, and the success of their cause. One of them said, in a very sarcastic tone, "McDowell is a fine general: why don't you give him sole command?" The absurd suggestion created shouts of derisive laughter, in which friends and foes heartily joined. Upon the walls of a building which was the headquarters of the brigade commander I read some inscriptions which had been written by the pickets of the enemy:

---

indicating the arrival of troops. I am returning to camp." Sickles wrote that Patterson "is quite ill" and should be relieved of command. Patterson committed suicide later that month. [OR XIX/2, 562]

10 At the time Blaisdell commanded a provisional brigade of the 11th and 16th Massachusetts and the 72nd New York. Sickles commended Blaisdell for "great zeal and activity" in posting guards on railroad bridges as they proceeded toward Warrenton. [Ibid. 562]

"Away goes the Yanks when they see the rebels approach them;" and, "T. W. Snead will never wear the galling yoke of a Northern Parliament."

The army began to move to Falmouth upon the 16th [of November].[11] The troops for two days were passing through this place; and general confidence concerning the result was expressed, and many asserted that Richmond would be captured within a month. This change of base[12] had not been anticipated; and workmen, who constructed two water-tanks at Bristoe Station upon the 18th, removed them upon the 19th. The regiment waded through the Occoquan, at Wolf Ford, upon the 21st; and the brigade occupied a position of great natural strength, which the enemy had fortified when Centreville was held; and forts had been built upon the crest of the hills, while the pines and cedars had been felled upon many acres to form the abattis. These earthworks were not considered perfect by the general [Carr]; and a detail was busily engaged in throwing up a new redoubt, when the orders to march were received. The brigade disappeared, and encamped in the midst of the short pines of Falmouth upon the 28th. A division general[13] discovered the skins of some sheep in a field in which his troops had bivouacked, but was unable to find any mutton or criminals, and arbitrarily deducted a supply for one day from the rations of his command as a punishment.

The shameful negligence of certain officials to forward the pontoons from Washington caused a fatal delay in the movement

---

11 Maj. Gen. Ambrose Burnside had just replaced McClellan as commander of the Army of the Potomac. about whom, somewhat surprisingly, Blake offers no comment.

12 From Manassas Junction, on the Orange & Alexandria Railroad to Belle Plain Landing on the lower Potomac.

13 Presumably Sickles.

of the army;[14] and Lee was enabled to mass his forces upon the heights in the rear of Fredericksburg, and fortify these strong positions, while Jackson started from the Valley, and reached the field the night before his men were required for action. The northern bank rose abruptly from the Rappahannock, and completely commanded the city, which was compactly built upon the opposite side; and the narrow stream flowed between the pickets of both armies, who gazed at each other from day to day without exchanging shots. Conversations were frequent until they were prohibited by the officers; and the following remark was often made: "Yanks, before you can take Fredericksburg, you will have to get up Early, go through a Longstreet, cross a Lee, jump over a Stonewall, and climb two Hills." Many wore the Federal uniform; and the rebel sentinels sometimes put on the overcoats of those that they relieved, and the reserve crept into the shelters and caves which had been excavated in the bank. The streets were filled with vehicles of all varieties, which were loaded with the baggage of the terror-stricken inhabitants, who were leaving the city to avoid the dangers of the battle that was looked for every day. Before the wharves at Acquia Creek had been completed, acres of ground near the stations were covered with army-wagons that occasionally waited three days for supplies; and the wheels were rumbling over the roads at all hours of the day and night.

Orders were received, in the evening of December 10, to furnish every man with sixty rounds of ammunition, and rations for three days; hospitals were established; and the soldiers beheld upon every side the usual preparations for a general engagement.[15] The reports of two cannons reverberating with a

---

14 Burnside wanted to cross the Rappahannock at Fredericksburg in mid-November and threaten a move toward Richmond.

15 Just prior to the battle of Fredericksburg, a sixth regiment was added to

peculiar distinctness in the darkness, at 5½, A.M., upon the 11th, broke the quietness of the camps; and the same rounds, succeeded by a volley of musketry, were heard fifteen minutes afterwards. These were the guns that opened the battle of Fredericksburg. Near the ruins of the railroad bridge, the engineers had built one of pontoons, which extended two-thirds of the distance across the river; but the canister of the enemy prevented them from finishing it at that time. The division marched at daybreak towards the point; and one hundred and forty-three pieces of artillery were placed in position upon the bluffs of the north bank, while most of the infantry was concealed in the woods and ravines. A dense fog, which prevailed during the morning and forenoon, rendered the progress of the general movement as hazardous as a conflict in the night; and the delay that occurred in laying the pontoons allowed the foe time to unite the troops that guarded the fords with the main body. The fire of the batteries upon the left, and the gallantry of the forlorn hope upon the right, triumphed over all obstacles; and the bridges at Deep Run were finished at noon, and those at Fredericksburg were completed three hours later. One hundred thousand infantry, and a force of cavalry and artillery, debouched from these two points of crossing, which were nearly three miles apart, and formed in line of battle. There was no fighting during the day between large bodies of troops; although the skirmishers were actively engaged, and the cannonading was sometimes brisk. The most deafening roar resounded when the guns opened upon the town with shot and shell; and clouds of smoke arose from burning edifices in every

---

Carr's brigade: the 11th New Jersey, Col. Robert McAllister. Burnside arranged the army into three Grand Divisions, each with two corps. III corps was part of the Center Grand Division, but its 1st and 2nd divisions were temporarily assigned to the Left Grand Division, which, with certain exceptions, was not heavily engaged at Fredericksburg.

district. The brick houses protected the rebel sharpshooters, who frequently attempted to deceive the Union forces by clothing themselves in the dresses which they found in deserted buildings.

The division was held in reserve upon the 12th; and from the field it occupied could be discerned the rifle-pits of the enemy upon the ridges, the national columns moving to their new positions,[16] and the batteries which successfully silenced those of the foe throughout the battle. The first cannon was discharged at 9.20, A.M.; and for half an hour a vigorous firing continued, in which the siege-guns planted upon the banks penetrated the innermost line of works, while they replied at long intervals with shells that could not reach the superior ordnance. The bivouac in the night affected the raw troops, that were constantly coughing; and new regiments could be quickly pointed out by this means. They were anxious to go into the fight, and "eager to enter the fray," in newspaper language; while the veterans, like old soldiers, did not wish to deploy upon the field unless their presence was indispensable. The division marched to the left, and halted for the night; but the regiment received orders for special service at 9½, P.M., and crossed the river upon a bridge that was composed of sixteen pontoons, which it was required to guard.[17] There were no fires, because there was no wood; and the men walked to and fro during many sombre hours to escape the chills that threatened them if they sought sleep.

A state of stillness which was unnatural, when the proximity of hostile armies is considered, [had] existed until mid-day of the 10th: but the dispositions preparatory to an attack upon the

---

16 Sickles' division (2/III) observed the artillery duel, and the crossing of the Rappahannock by troops of I and VI corps.

17 The bridge guard consisted only of the 11th and 16th Massachusetts, 2nd New Hampshire, and 11th New Jersey. The 1st Massachusetts and 26th Pennsylvania remained with the division.

enemy had been made; and the history of that afternoon [December 13th] would be aptly written in the blood of the gallant soldiers who assaulted impregnable works upon the right, and defied death at the stone wall, the heights, and the mill-race. A citizen said with truth, to some companies that were marching through the city to the front, "The soldiers upon those hills are looking down, and laughing to see you advancing to meet them." The lines extended from Deep Run a mile and a half to the left, and deployed upon the plateau that was nearly two miles from the river; and the main portion of the army, comprising eight divisions of infantry, with 60,000 men, 28 batteries containing 116 guns, and the force of cavalry which was sheltered by the bank in the rear, awaited the orders of Gen. Franklin. The enemy was hidden in the belt of woods in front; but the land was very level: and there were more of the insurmountable obstacles that blocked the path to victory upon the right; and a competent commander with such a large corps would have easily carried the position, which was defended by troops that were inferior in numbers and resources.[18] When the opinions and sympathies which this officer entertained upon the vital issues of the Rebellion are publicly known, and his inglorious military career, from first Bull Run to the disgraceful failures of the Sabine Pass and the Red-River Expedition, is scrutinized, all will be amazed that Gen. Franklin was entrusted by the Government with any command in the service.[19] The batteries shelled the forest; and at

---

18 The performance of Franklin and Burnside at Fredericksburg is an unending controversy. Franklin's attack was half-hearted; he claimed his orders were not explicit. Burnside blamed Franklin for the failure of his questionable frontal attacks on Marye's Heights.

19 The relationship between Franklin and Burnside festered until the latter was replaced by Hooker in early 1863. Franklin was transferred to command of XIX Corps, which he led in the Red River expedition of 1864. He was wounded at the battle of Sabine Crossroads, Louisiana, April 8,

one, P.M., the ceaseless roll of musketry burst forth for the first time during the movement, and a single division gained a temporary success: but the inexcusable neglect to support it, resulted, as a matter of course, in defeat. The troops of one of the largest corps, with the exception of a few skirmishers, did not burn a cartridge; and less than one-fifth of this vast army upon the left fought in the decisive battle.[20]

The wounded who could travel were continually returning from the front, and the helpless were carried by the ambulances and stretcher-bearers over the bridge; and great vigilance was necessary to detect those that feigned sickness or wounds by tying a bandage stained with blood around their arms or heads, and prevent them from escaping across the river. An officer, assisted by two able-bodied men, slowly moved towards the bridge, until the colonel [Blaisdell] halted them, and directed the soldiers, in terms of the deepest kindness, to rejoin their company, and assured them that their commander should receive the best treatment.

"My good man, what is the matter with you?" he blandly asked the lieutenant who had requested that those who bore him from the field might be allowed to remain and assist him.

"I am wounded," he replied in a weak voice; and an expression of the most acute pain was visible in his face.

"Doctor, will you dress his wound? He is just from the front."

1864.

20 Reynolds (I Corps) attacked about 8:30 A.M. on the 13th, with Meade's 1/I division achieving temporary success. Gibbon's 2/I division was the immediate support, but lost contact with Meade in the woods, and retired. Birney's 1/III division was also sent to support. Blake laments that the entire III corps was not sent in.

"I didn't say I was wounded: I am sick, and want to go over the river to be treated by my own doctor," he said when he saw the surgeon approach.

"You can go as soon as you have been examined and reported unwell."

"I will go any way," the officer exclaimed, and tried to rush by the guards who arrested him. The colonel changed his soft words into hard oaths, struck him, and ordered the men to use the bayonet if he resisted them; and the skulker ran towards the front without showing any loss of physical strength. Many scenes like this occurred during the afternoon; and the exact situation of affairs was ascertained from the disabled, who were always willing to tell the news in answer to the usual question, "How are things going?" When the facts attending the death of Gen. Bayard[21] were received, the soldiers publicly uttered the wish that the cruel shell had missed its noble victim, and pierced Gen. Franklin, who was standing near him at the time. The [rebel] prisoners were happy because they supposed that their lives were safe for a certain period; and one of them remarked, when he saw a group of mounted men riding upon the distant heights, "That is Longstreet upon the white horse, and his staff."

The regiment rejoined the division at midnight upon the plateau,[22] and learned the position of the enemy by watching the lights of the camp-fires, which shone with distinctness in the darkness. The skirmishers commenced to fire with the first ray of sunlight upon the 14th: and until one, P.M., the sharpshooters, who were posted in the woods about a quarter of a mile from the

---

21 Brig. Gen. George D. Bayard, commander of the cavalry brigade assigned to the Left Grand Division, was hit by a random shell fragment at Franklin's headquarters. [Sifakis, 25]

22 On the night of December 13-14, the 11th Massachusetts and 2nd New Hampshire were relieved of their bridge-guarding duty, followed by the 11th New Jersey and 16th Massachusetts the next morning.

line, shot at every person and horse that stood upon the plain, and occasionally wounded a man; and soldiers who were aligned three hundred yards in rear of the regiment were killed by the bullets which whistled over it. The field had been planted with corn; and the beds were made of shucks and stalks, that were collected together, and placed in little gulleys and ditches which ran through it. The rebels had burned a house and barn in the night that interfered with the range of their artillery. A [rebel] battery opened on the brigade at daybreak, but it was promptly silenced by a company of sharpshooters from the 2d New Hampshire Vols.; and a few cannon and officers, drivers, and horses, fled in confusion, and left the guns and caissons. Throughout this contest,[23] the skirmishers sheltered themselves behind stumps and other barriers; and some scooped up a slight quantity of earth, and rested their rifles upon the bodies of dead soldiers that were frequently mutilated by the balls which were aimed at the living.[24] A lieutenant in the brigade was wounded in the extreme front, and refused to allow the man to carry him to the hospital while the firing continued; and rejoined his regiment within a month, before his injuries were healed, when there was a prospect of another battle.

"Captain, where shall I bring your dinner?" asked a servant who was retiring to cook that meal.

---

23 Company B, 2nd New Hampshire, made the foray against the rebel battery: "The skirmish was kept up until near night, the Second keeping out one company at this time, relieving as fast as ammunition was exhausted." [Haynes, 148-149]

24 Sickles reported that "the sharpshooters and skirmishers of the enemy, covered by the wood in front and some farm buildings and an orchard on the right, kept up a constant fire, from which the troops on the right suffered most. To this I could only oppose the fire of my own skirmishers, who were much exposed." [OR XXI/1, 379]

"I don't know: in hell, perhaps!" the officer[25] answered as he glanced at a shell which burst near the spot at that moment.

The firing ceased in the afternoon, and a tacit truce existed, during which the rebels permitted the members of the ambulance corps to convey the wounded to the rear, and brought others who were inside of their lines to the edge of the woods which was neutral ground.[26] A rare spectacle in war was witnessed when the soldiers of both armies talked together in the most friendly manner upon this space between the pickets, while an animated conflict could be distinguished at the distance of two miles to the right. They wished to exchange tobacco for "picture-papers," because the ordinary news-journals did not interest the large majority, who were unable to read. They informed the men that Jackson commanded the army in front; and said, "In three days he will drive you into the river, or make you cross it." A private noticed a rebel officer, who was a native of the same town, and lived near his home until he emigrated to the South two years before the Rebellion: but, the instant that he recognized him, a feeling of utter degradation seemed to overcome him; and, without saying a word, he rushed into the forest to conceal his emotion of shame.

"Good-bye, boys; we will meet you in the fight to-morrow," one of the regiment remarked when night approached; and the crowd dispersed. The rebels easily counted the guns and troops, which were massed upon the open plain, and formed the left of the army; and, wisely fearing the results of an advance by these

---

25 Presumably Capt. Munroe of Blake's own Company K.

26 Sickles noted: "the stretcher-men from my ambulance corps, … occasionally went unmolested to the verge of the enemy's line to get the wounded of Gibbon's division who fell on Saturday [the 13th]. The stretcher-men were told by the enemy that if our skirmishers would not fire any more, our ambulance parties might come anywhere along or within their lines and get all of our wounded." [OR XXI/1, 379]

battalions if they were properly handled, worked after sunset on the 13th, 14th, and 15th, and felled trees and erected breastworks until morning. One-third of the men in each company [of the 11th Massachusetts] were kept under arms during the night; and the only sounds that fell upon the ears of the faithful pickets were the strokes of a thousand axes, and the crash of the massive oaks when they struck the earth.[27]

The same good feeling prevailed upon the 15th;[28] and, although the troops might be exposed to the fire of the artillery, the Union forces dug no extended rifle-pits, the brigade was never ordered to load, and the foe showed no wish to molest them. The regiment was relieved for a part of the day,[29] and remained in the old road, the grade of which was several feet below the surface of the plateau, and reminded one of the famous sunken road of Waterloo. A flag of truce entered the lines in the afternoon;[30] and the gunners, who had fled from the battery near the burnt chimney at sunrise upon the 14th, attempted to take away the caissons and cannons which had remained in the same position more than thirty hours. The watchful captain opened his artillery upon them, and they again, within a few minutes, deserted their pieces; while dismounted officers and men and riderless horses scampered with a speed that caused general laughter. A company from the regiment was stationed upon the picket, and an agreement was made that there should be no

---

27 Carr stated that on the night of December 14, the 11th Massachusetts was "engaged in throwing up rifle-pits at the front to protect our skirmishers." [OR XXI/1, 383]

28 Original text say "18th."

29 Carr wrote that his 1/2/III brigade was relieved at 10:00 A.M. on the 15th by the 3/2/II brigade, and then at 6:00 P.M. returned to relieve the 2/2/III brigade. At 9:30 P.M. they got their orders to retire. [Ibid. 384]

30 This must have been a local event, since it is not mentioned in any official reports.

firing; and the enemy began to labor in the night with unusual industry. The sentinels discovered, by noticing the clanking iron, rumbling wheels, and similar sounds, that cannon were being mounted upon the forts to sweep the plateau, which presented no natural or artificial obstacles which would check the passage of shells and canister. Dippers were placed in the haversacks to prevent the incessant tinkling which is always made by marching troops; the force quietly glided from the fort a few minutes before midnight; and the pontoons creaked beneath the tread of a discomfited army during the succeeding four hours. The weather, which is a subject of stale conversation in peace, but of the greatest importance to soldiers in a campaign, had been pleasant up to this time: the ground was hard, and the heavy guns rolled over the roads with ease. Nature now changed its kind aspect to favor the retreat; overcast the skies with black clouds that shut out the light of the moon, so that the hostile forces could not see; and roughly waved the branches of the forest over their heads, so that they were unable to hear; and the divisions escaped to the old camps undisturbed by a solitary shot. The Union pickets did not know the time when the main body marched to the river, and cautiously crept upon their hands and knees to the sunken road, when they received the orders to fall back.[31]

The Virginia Central Railroad passed through the rebel lines, and trains constantly conveyed ammunition and re-enforcements from Richmond, or carried the wounded and prisoners from the battle-field; and the smoke that arose from the conflict mingled with that of the locomotives. The sutlers, and storekeepers of Fredericksburg, concealed large quantities of tobacco; and the soldiers, among whom there was always a senseless clamor for a

---

31 Sickles wrote: "at 10:30 I moved by the flank across the river…all the troops, except the skirmishers and supports, crossing over before midnight." [OR XXI/1, 381]

"chaw" or "smoke" (I have seen fools barter a day's ration of bread for a small piece of the weed), eagerly obtained a supply from some boxes that were scattered upon the bottom of the streams. Some of the troops that bivouacked within the limits of the city pillaged the deserted houses of rich rebels who had cheerfully allowed the sharpshooters to fire from the windows and murder their comrades. Mirrors, pianos, and gorgeous furniture were destroyed; beautiful paintings and family-portraits upon the walls were cut; busts were decapitated; and elegant silk dresses and garments were torn into shreds. The bricks which fell from the chimneys during the bombardment, and the partitions of shattered buildings, injured the soldiers who were deployed in the streets; while the inhabitants that had not escaped crouched in cellars, and dreaded alike the balls that came from friends and foes. A citizen who viewed the subject from a personal stand-point, in commenting upon the conduct of the army, said, "All soldiers are the same: the Confederates robbed me of all I had, and you Yankees took all I had left."

The rain gushed from the clouds for hours upon the 16th; and, if there had been any delay in recrossing the river, the cannons and wagons would have been fixed in the adhesive mud of a Southern winter, and the most disastrous results would have followed. The forces of the enemy advanced in line of battle in the morning as soon as the evacuation was perceived, plundered the dead, and gathered all the clothing that had been cast aside by the army; and a battery opened upon them, and the battle of Fredericksburg was finished.[32] The repulse caused universal despondency; and the soldiers of Lee exultingly told the detail that crossed the river under a flag of truce to bury the dead, that

---

32 Sickles' 2/III division sustained just 100 casualties in heavy skirmishing. On the other hand, Birney's division (1/III) lost nearly a thousand supporting Meade and Gibbon.

there would be no more fighting, and the Southern Confederacy would be acknowledged as a nation within two months. The rebel generals urged their hordes in each conflict to win the victory, and then they could return to their homes, and enjoy the rights for which they were contending, and — *Reap the harvest of perpetual peace By this one bloody trial of sharp war.*[33]

---

33 *Richard III* Act V, Scene ii (Shakespeare).

Chapter IX

The Camp at Falmouth, Virginia

FOR THREE DAYS the army had been under marching orders; and the division, equipped for battle and commanded by Gen. Sickles, moved two miles at one, P.M., upon January 20, 1863;[1] halted until night in the severe storm, and then returned to quarters. The troops that belonged to the left grand division,[2] under Gen. Franklin, filed by the regiment, and muttered bitter curses against certain officers who exhausted them by moving at an unnatural rate of speed, and giving no permission to rest, although most of them were eight miles from the starting-point. These sinister commanders effected their purpose, and filled the woods that bordered upon the roads with thousands of stragglers who could not sustain the cruel fatigue of the march; and some reckless men openly insulted them, and shouted defiantly as they brandished their bayonets, "Shoot us if you want to kill us!" "Ride over me if you dare to!" "Get off your horse, and carry this

---

1 Blake omits discussion of the five weeks after Fredericksburg for a very good reason: he was absent without leave for much of the time. In his memoirs he wrote of his younger brother William, who also served in Company K, that "his constitution was undermined by the destructive exhaustion of campaign" and that "he was a wreck" when they arrived at Falmouth. After William died in the brigade hospital, Blake recalled his furtive journey back to Boston with his brother's body:

"There was a probability of a winter campaign and I applied for a leave of absence of ten days to enable me to transact necessary business and attend the funeral. This was refused peremptorily at headquarters, but I was granted two days within which to go to Washington and return to camp. Col. Blaisdell notified me of the result and indulged in rough words uttered with a good heart, saying it was an outrage and I was a fine officer and entitled to be absent, and made this dangerous suggestion: 'You go to Washington with your leave but don't stay over ten days; look after the shipment of your brother and take the cars for Boston. I will report you absent with leave, and if you hear there may be a movement at the front,

knapsack if you like it!" while hundreds screamed, "Halt, halt, halt!" or, "I'm demoralized!" Less than one-half of Gen. Franklin's force that shouldered muskets in the afternoon formed a part of the ranks when the lines were established at night. This dishonorable conduct of certain officers of high rank, who did not cordially support Gen. Burnside, produced among the soldiers a feeling of distrust regarding the success of the movement before the river was bridged.[3]

The rain continued hour after hour, while the division struggled through the mud, upon the 21st, for a distance of five miles to the right, and halted at a point that was near [Bank's] ford at which the troops were to cross the Rappahannock. The soil of Virginia was a more formidable obstacle than the legions of Jackson and Longstreet. The animals were constantly mired, and four mules were entirely exhausted by drawing nine hundred pounds of rations from the camps in a wagon; while twelve horses attached to a light cannon extricated it with difficulty, although four of them pulled the gun upon ordinary roads. When they bivouacked, logs were placed under the wheels of the

---

get back as fast as you can. You're smart enough to get through somehow, and I will look after you here.'

"A large congregation attended the funeral at the Harrison Square Church. A week flew, and I glanced one afternoon at the bulletin board before the office of the *Boston Journal* and read 'Rumored advance of the Army of the Potomac.' I bought newspapers, scrutinized the telegrams, got my railroad ticket, went home and said 'good bye,' and was off on the first train to the south. I pushed ahead, lost no time and joined my company while marching in the so-called 'Burnside Mud Scrape.' Neither I nor Col. Blaisdell had any trouble through the adventure, but we incurred the risk of a Court Martial and punishment which might be dismissal. However, the spirit of favoritism pervaded the service, and fighting officers like Col. Blaisdell were indulged like the best scholars at school, and could not be spared from the battlefield."

2 Referring to Burnside's organization of the army into Left, Center, and Right Grand Divisions consisting of two corps each.

3 How active other senior officers were in undermining Burnside's second

artillery and teams to prevent them from sinking into the earth which wished to receive them; and most of the pontoon train was firmly planted at different points. The rebel pickets, who understood the state of facts in the army, expressed their delight by performing somersets and other feats of agility, and shouted, "Why don't you cross the river?" "Bring up the pontoons, and we'll lay them for you." "The Yanks are stuck in the mud." The campaign ("mud-scrape") was necessarily abandoned; the soldiers corduroyed the roads during the next two days with fence-rails and trees, so that the batteries could safely return; and the whole army followed them to the old camps, and began to erect winter quarters.

The division, enveloped in the flakes of falling snow, marched to Hartwood Church upon February 5 to guard the fords upon the river; while a force of cavalry advanced to Rappahannock Station, and burned the bridge which the enemy had recently constructed. The brigade and a battery were posted near Richard's and United-States Fords; and at the last-named place the short pines had been felled to put the artillery in position; telegraph-poles had been erected, and the way had been cleared for the passage of troops. The foe built some redoubts upon the opposite bank, to command the point of crossing, a few hours before the plan of attack was made; and Gen. Burnside was again baffled. The trees had been marked to guide the cavalry to various positions; and many pines in the forest were snapped asunder by the weight of snow. The regiment passed by the "gold mine," upon which operations had been suspended since the beginning of the war;[4] and a woman, who stood in the doorway

---

campaign, which they opposed, is not clear. Pro-McClellan generals Franklin and Smith had, following Fredericksburg, proposed a return to the Peninsula instead of another attack across the Rappahannock.

4 One of several small-scale gold and iron mines in the Wilderness area

of a house, said, "If I was a man, I should shoulder a musket, and shoot some of you." A squad of the butternut cavalry was upon picket; but the river was not fordable at this time, and the division rejoined the corps as soon as the force returned from Rappahannock Station.

While serving upon the [brigade] staff during this expedition, I witnessed an incident which illustrates military matters in certain respects. The major of a squadron of cavalry casually remarked that he was born in the state of Rhode Island; at which a brigade commander [Carr] spoke, "One of my best officers came from Rhode Island." He then introduced to [the major] a chaplain, and added, "And a d—d good chaplain he is too." The clerical subject of the conversation, with a smile of satisfaction, thanked the person that uttered the compliment. The general, who lived in a comfortable brick mansion that was four miles from a part of his division and eight miles from the remainder, never visited the fords which he was ordered to protect; while the troops, exposed to the merciless sleet, were stationed at their posts to resist the attack that was expected. However, he issued a pompous order of thanks to the soldiers that he had seen upon two occasions when they were marching from and to the camps, commended in tender words their fortitude in enduring the storm, and declared that he felt a pride in leading such brave men to scenes of danger.

The winter season was marked by no unusual features, and the routine of camp and picket guard, and labor upon the roads, constituted the military duties for four months. The enormous quantity of wood that was required for the barracks and camp-fires rapidly exhausted the forests, and thousands of acres were covered with stumps. An old resident said to the pickets that were

____

during the mid-19th century.

posted near his house, "After the war there will be no rails, and no wood to make them." Buildings were leveled; fences burned; the bricks and stones of capacious chimneys formed the flues of log huts; the wagon trains and batteries cut new thoroughfares across estates; the feet of men, and hoofs of horses and mules, trampled fields of vegetation into barren wastes; every landmark was destroyed; and the work of destruction within the lines was complete.

The indolence of soldiers dwelling in a permanent camp, when their efforts are not stimulated by the presence of an enemy, is remarkable; and ten men from a new regiment will perform as much labor as one hundred veterans, until they have been corrupted by the bad example of shirking comrades. Upon March 4th, four hundred and fifty soldiers, supposed to be working seven hours, corduroyed one hundred and fifty feet of an old road that was thirteen feet wide, although the logs were [already] cut upon the bank, and covered it with a slight quantity of earth. There is much more grumbling and discussion in the camp than on the march: the mind is active when the body is inert; and it is a singular fact, that those whose physical comforts were the least at home make the loudest complaints about the quality of the rations and the Government clothing; and the men that earned a livelihood with the shovel [in civilian life] were generally the most unwilling to handle it in the army.

The soldiers who had fought under Gen. Hooker were delighted when he was appointed to command the army;[5] but many officers of high rank were dissatisfied, and, assisted by others who had been most justly deprived of their positions,

---

5   Franklin and Smith expected that their efforts to remove Burnside would pave the way for the return of McClellan. But radical Republicans in the administration would not stand for this, and Hooker was chosen for political reasons [Sears 1996, 22-24], assuming command of the army on January 26, 1863.

sought to undermine the confidence of the people and enlisted men by representing him as a common drunkard.

The cavalry, which had hitherto rendered little if any actual service, and was usually detailed for ornamental and escort duty, was inspired with a new energy when he re-organized this invaluable auxiliary,[6] and ordered the officers to prepare for the severe fighting from which they had been so long exempt. A vast injury had been inflicted upon the country by the [War] department which followed the erroneous advice of Generals McDowell, Franklin, and others who entertained the same views, and prohibited in the first year of the war the formation of the mounted regiments that were demanded. The following brief extracts are taken from their testimony under oath before the Congressional Committee, and show a deep ignorance of the science of war, which seems astounding when their advantages of education at West Point are considered.[7] Gen. Franklin said, in different answers, "We have a great deal more cavalry than any of us need." "I really think that two thousand cavalry is all that we want for the whole army. I would not give a snap for more than one-third of what I have now." He had less than a thousand cavalry with his division, that exceeded twelve thousand soldiers. "I would never think of making a cavalry fight or a cavalry charge." Gen. McDowell remarked in reply to questions, "I think we might do with less than that" (meaning thereby less than twelve regiments of cavalry to one hundred and forty-four of

---

6 Under Hooker, the cavalry of the Army of the Potomac was consolidated into a single corps, although it operated as three separate divisions most of the time until 1864. Before then, Union cavalry typically was used in forces no larger than a regiment.

7 "In an army which in 1861 had only two serving officers (both in their seventies) who had ever commanded in the field a military unit as large as a brigade, no one had any firm grasp on … what the primary mission of the cavalry in a modern war should be." [Starr/1, 60]

infantry). "If we were to be organized by corps of three divisions each, two regiments of cavalry would be perfectly sufficient for the three divisions" (meaning thereby that a force of eight thousand cavalry "would be perfectly sufficient" for an army of one hundred and forty-four thousand infantry).

The world has long ridiculed Phormio, the civilian, for delivering an oration to Hannibal upon the strategy of war;[8] but he [Phormio] was wiser than these military teachers who were examined by Congressmen searching for knowledge. A private who should express shallow opinions like the foregoing would not be recommended for the rank of corporal in a colored regiment by the competent board, that "would not give a snap for more than one third" of such an ignoramus. What direful results would have ensued if these generals had succeeded in their aspirations to attain the command of the Army of the Potomac, and paralyzed that cavalry which has won a glory as brilliant as that of the divisions that followed the white plume of Murat![9] A force of cavalry crossed the river at Kelly's Ford upon March 17; defeated the enemy in a desperate fight and charge, which amazed those who "would never think" of making them; and returned to the camps elated by their first victory upon the field of battle.[10]

---

8 Hannibal once listened to a lecture by the philosopher Phormio on the duties of a general. When asked his opinion, he replied, "I have seen during my life many old fools; but this one beats them all."

9 Joachim Murat, French cavalry commander. At Eylau, February 8, 1807, Murat ordered all of Napoleon's cavalry to attack the Russian center in an attack that went down in history as the "charge of the 80 squadrons."

10 "The 2nd Division of the Cavalry Corps, under Brigadier General William Averell, crossed the Rappahannock at Kelly's Ford on the morning of March 17, 1863. The main action developed about three-quarters of a mile from the ford: through a series of charges and counter-charges Averell was able to push the Rebels back a couple of miles." [Union Army, Vol. 6]

Gen. Kearney, at Fair Oaks, ordered his soldiers to sew a piece of red flannel to their caps, so that he could recognize them in the tumult of a battle, and detect those who attempted to evade the performance of their duty. Gen. Hooker introduced into the whole army the system of badges, which was of incalculable value; and henceforth the members of the various divisions in the corps wore, as emblems of honor, the red, white, or blue circle, the trefoil, the lozenge, the Maltese and plain cross, and the star. The men inscribed upon them the names of generals whose memories were cherished; and, disregarding the actual commander, styled themselves as soldiers of their "old brigade," or "old division."[11]

Thus month after month passed quietly away. The grass began to sprout in the sods of the barracks, and the rumors of an advance daily thickened. No shots were exchanged between the pickets, who pushed their bayonets into the ground, and made water-wheels or ornaments of bone and laurel-wood to occupy the weary hours, and floated across the river boats and sticks, to which newspapers were fastened for sails. The brigade guarded the right of the army, and remained three days upon this duty, during which it was usually commanded by a field-officer, while the general enjoyed the safety and comfort of his tent in the camp. Upon one occasion, when Stuart crossed the stream and an attack was anticipated, a brigadier boldly ordered a colonel, who had arrived from his home which he had visited with leave, to proceed at once to the front, and take charge of the troops. The conduct of this starred poltroon[12] was in striking contrast with

---

11 Hooker's Chief of Staff, Maj. Gen. Dan Butterfield created badges for each corps. Kearny had invented the idea for his own division the year before, so the III corps retained the diamond badge. The men of the Carr's Brigade, as part of the second division, wore the white diamond.

12 In all likelihood referring again to Carr.

that of commanders like Generals Kearney, Hooker, Grover, and others, who frequently inspected their picket-lines, and bivouacked at the front whenever a contest was imminent. It was common practice to fish in the Rappahannock, until it was forbidden, because the citizens of Falmouth were detected in the act of signalizing to the enemy by means of the poles.[13] The soldiers generously supported a foe, who was a cripple, and lived with his family near the outposts; and every relief shared with him the rations, split and carried wood to his hovel, and heaped coals of fire upon his hearth.

The removal of the surplus stores of private and public property, and the activity which was visible in every branch of the service, were the forerunners of another campaign; and the disappearance of the mud would be the signal for the forward march. Inspections and reviews were often witnessed; and the spectators beheld with admiration the generals, with their gayly dressed staffs, mounted upon richly caparisoned steeds, as they dashed with great speed along the extensive lines, and the martial tread and evolutions of the vast columns comprising 60,000 soldiers upon the cheerless plains of Falmouth. These occasions were extremely distasteful to the rank and file, who viewed with indignation hundreds of showy officers who were non-combatants, and never faced the rebels upon the battle-field, but lurked in the rear with the trains of wagons and herds of cattle.[14] I have frequently seen only four or five aides with the general

---

13 Hooker issued orders prohibiting communication between opposing pickets across the river. For more on Hooker's security measures see Fishel, Chapters 12-16.

14 Despite the presence of craven parade-ground officers, the morale of the army soared in early of 1863. In addition to seeing that the men were properly sheltered and fed, Hooker clamped down on desertion. Because many deserters received civilian clothes in the mail, Hooker ordered all parcels inspected. [Sears 1996, 76]

when the conflict was raging, while upon the harmless parade the number increased to a score, who rushed over the ground with the velocity and importance of a "great god of war," in a militia sham-fight. Orders were received upon April 13 to march with rations for five days in the knapsacks, besides the ordinary supply for three days in the haversacks; but a flood checked the movement on the day that was designated,[15] and the food was consumed in the camps. I quote, in this connection, another strange and inaccurate statement from the pamphlet of the Prince de Joinville, without further comment: "In Europe, our military administration assumes that the transportation-service of an army of one hundred thousand men can only provision that army for a three-days' march from its base of operations. In America, this limit must be reduced to a single day. An American army, therefore, cannot remove itself more than *one day's* march from the railway or water-course by which it is supplied."[16]

---

15 The flooding on April 15 delayed the cavalry corps crossings at Kelly's Ford and Rappahannock Station.

16 The "pamphlet" is Joinville's *The Army of the Potomac: Its Organization, Its Commanders, Its Campaigns.* Blake offers no comment because it was routine for troops to start a march with rations for three-days.

Chapter X

The Battle of Chancellorsville

THE DIVISIONS of the army, laden with sixty rounds of cartridges, and rations for eight days, comprising five articles, — salt pork, hard bread, salt, sugar, and coffee,[1] — broke up the encampments upon April 27, 1863, in which they had lived for the period of six months; and quit them for the fourth time, with the expectation that they would never return to Falmouth. The long columns of infantry; followed by the batteries, the packed mules, and the ambulances, marched in the afternoon a tortuous course behind the hills upon the northern bank of the river to conceal their movements from the eyes of the enemy.[2] At midnight, three corps, the first, third, and sixth, concentrated at the same point, near Deep Run, which the regiment held in December; while the remaining corps, including a part of the second, passed by the brigade when it was upon picket, and moved in the direction of the upper fords of the Rappahannock. The pontoons were placed upon the shoulders of the men, who transported them to the river, over which they stealthily glided, and captured the rebel force that was sheltered behind two rifle-pits to resist the attempt of any body of troops to cross. Two divisions were deployed in line of battle upon the opposite bank during the day:[3] the hostile skirmishers sat upon the plain a few

---

1 The standard marching kit was three days rations and forty rounds of ammunition. Hooker was testing a method of longer duration marches requiring less dependence on supply trains. For Chancellorsville, each man carried over 40 pounds of arms, food, and equipment. [Sears 1996, 104-106]

2 The march route kept well north of the Rappahannock, and no fires were permitted overnight April 27/28.

3 On April 29, pontoon bridges were placed at Franklin's Crossing for

rods apart, without firing or making any advance; and, at times, the flashing bayonets showed that the columns of the enemy were marching in the sunken road. It rained at intervals upon the 29th, but no changes were observed in the dispositions upon the left; and these troops, exposed to cannon and thousands of rifles, which might open upon them at any instant, stacked their guns, pitched shelter-tents, cooked their meals, and the officers of many companies, upon both sides of the river, were making the regular muster and pay-rolls.[4] A dense fog obscured the view during the forenoon of the 30th; but the same state of quietness reigned until the order of Gen. Hooker was read to the soldiers amidst loud cheers, when they learned that the foe must "ingloriously fly," or "give us battle on our own ground, where certain destruction awaits him."[5] The balloon rose at noon in the south-west, and appeared, from this standpoint, to be in the rear of the Heights of Fredericksburg, and removed every doubt regarding the success of the Union arms. A few minutes before twelve, M., [April 30] the division, commanded by Gen. Berry, and the third corps, under Gen. Sickles, meandered [fifteen miles] through valleys and over the slopes of rising ground; and no men were allowed to stand or walk upon the crest: so that they were shielded from observation. They bivouacked near Hartwood

----

Brooks' division (1/VI), and at Fitzhugh's Crossing, two miles down-river, for Wadsworth's (1/I). The same day V, XI, and XII crossed at Kelly's Ford, as did Stoneman's Cavalry Corps.

4   I and VI Corps, just below Fredericksburg and in full view of Lee, tried to present a semblance or normal camp activity.

5   Hooker's General Order No. 47, April 30, 1863: "It is with heartfelt satisfaction the commanding general announces to the army that the operations of the last three days have determined that our enemy must either ingloriously fly, or come out from behind his defenses and give us battle on our own ground, where certain destruction awaits him. The operations of the Fifth, Eleventh, and Twelfth Corps have been a succession of splendid achievements." [OR XXV/1, 171]

Church at midnight. The excessive weight which was carried by the troops increased the severity of the long march, the line of which could be traced by glancing at the overcoats and blankets which were scattered to alleviate suffering, and formed, in many spots, a carpet upon the road.[6]

The corps was in motion at the dawn of May Day; and reeled across the stream at United-States Ford, with the uncertain steps of the drunkard, upon the pontoon bridge which swayed to and fro in the rapidly-flowing current. One hundred squalid prisoners, who were captured at Germania Ford upon the Rapidan, passed by the column under guard; and one of them shouted, "You will need three years' rations before you can get Richmond." Their action in surrendering to the national forces refuted their constant boasts that they would fight until the last drop of blood was shed. The regiment rested in an old rebel camp until five, P.M., when the cannonading was active: the division was ordered to double-quick to the front, and formed at sunset a part of the army, of which the principal portion had been massed at Chancellorsville, with scarcely any loss, by the consummate ability of its commander.[7] The aristocratic slave-holders of Virginia, adopting the custom which their ancestors had brought from England, gave to their estates the name of the family, with the suffix of a -ly, or -dale, or -ville; and some of these domains, like Chantilly, had been the scenes of deadly conflict, and

---

6   Despite orders that each man was to carry only a single change of shirt, drawers, and socks (per the eight-day rationing plan), it seems that many over-packed.

7   On May 1 Hooker sent XII corps down the Plank Road and Sykes' (2/V) down the Orange Turnpike, while 1/V and 3/V divisions were ordered down the River Road toward Bank's Ford. When substantial opposition was encountered on these roads, particularly by Sykes, the forces were withdrawn and Hooker decided to form a perimeter around Chancellorsville.

become endeared in many Northern homes. Chancellorsville consisted of a large brick building, built in the style of the last century, and with the exception of massive pillars in its front, that extended from the basement to the roof, was very plain in its appearance. Negro cabins, cooking-houses, and other small out-buildings, were upon the grounds near the dwelling, which was occupied by the Chancellor Family; and the garden contained a private cemetery, which was planted with pines and savins, beneath the branches of which the kindred of the proprietors slept in their graves. The regiment loaded the muskets while it was moving upon the road which led from the United-States Ford; bivouacked in the forest, a short distance from the mansion; and the cannonading and picket-firing caused the formation of the lines for action several times during the night. Three companies were detailed to guard prisoners and cattle, and those who had charge of the latter found that it was the most difficult task.

The sky displayed its clearest blue at day-break upon May 2; and from right to left the skirmishers and sharpshooters of both armies were continually engaged; and — *Twixt host and host but narrow space was left, — A dreadful interval.*[8]

In the vicinity of Chancellor House there was a large tract of open ground, upon which the artillery was posted in large force; and the brigade, relieved of its knapsacks and haversacks, which were left in the woods under a small guard, marched to this point, and supported a battery which was planted across the plank road that ran from Fredericksburg to Orange Court House. The gunners cheered during a spirited contest with the enemy; and the huge volume of smoke which slowly ascended showed that a well-directed shot had exploded a [rebel] caisson. The troops, at this time, were in an excellent state of discipline: there were not

---

8 *Paradise Lost*, Book 4 (Milton).

six soldiers absent without leave from the regiment; and the few stragglers were arrested by the provost-guard, and placed with the skirmishers in the extreme front; and some received the mortal wounds which they had tried to shun by dishonorable conduct. The first line was resting behind a strong breastwork; and the soldiers were cooking coffee — the chief luxury of the army — upon their fires, as unconcerned as if they were in camp.

The regiment was detached for a special service[9] at half past seven, A.M., and advanced towards the run, with five companies deployed as skirmishers, upon both sides of the plank road,[10] in the form of an inverted "V" with the apex in the center of it, and forced back the enemy half a mile, until the muskets commanded the ground; upon which I saw a confused mass of dead horses, broken wheels, and the fragments of the shattered caisson. At three, P.M., the pickets upon the right heard a few sentences of the speech of a [rebel] officer to his command, in which he reminded the soldiers that the "Yanks" had plenty of rations; and concluded by asking, — "Men, will you sleep upon that hill to-

---

9   Sickles wrote: "I was directed to make a reconnaissance in front to the left of Chancellorsville. Major-General Berry was requested to detail two reliable regiments, led by circumspect and intrepid commanders. The Eleventh Massachusetts, Colonel William Blaisdell commanding, moving out to the left, toward Tabernacle Church, and the Twenty-sixth Pennsylvania, Colonel B. C. Tilghman commanding, in front, gallantly pressed back the enemy's pickets and skirmishers until he was discovered in force. A detachment of Berdan's sharpshooters, from Whipple's division accompanied each regiment." [OR XXV/1, 385-386]

10 The report of Capt. John S. Poland, Chief of Staff for Berry's division indicates that the 26th Pennsylvania advanced up the Plank Road and: "a reconnaissance was ordered to be made on the Banks' Ford road, running toward the left, for which Col. William Blaisdell, with the Eleventh Massachusetts Infantry, was detailed. Colonel Blaisdell, having taken a position exceedingly annoying to the enemy, received an attack of a rebel brigade to dislodge him, which his regiment gallantly repulsed. He communicated much reliable information, and for the services rendered received the commendations of General Hancock, commanding the lines to the left of the Chancellor house." [Ibid. 449]

night?" "Yes!" "Yes!" "We will!" "We will!" many voices answered. The skirmishers were quickly driven in by the onset of a brigade that advanced with fixed bayonets and a terrible yell that defined with accuracy the extent of the line of battle, so that the men knew in what direction to aim, — a fact of vital importance, that could not be ascertained in the forest, which was overgrown with stunted oaks. The companies that had been stationed upon the outposts hurriedly formed upon the reserve, which was posted in the road; and a battery enfiladed the ranks, while a fierce charge was made upon them by an infantry force. The enemy was repulsed[11] at the end of three-quarters of an hour in the most signal manner; and the commander[12] expressed his satisfaction with the glorious result by shouting, — "Now, my good men, give three cheers for hell! Three cheers for hell!" he repeated; and the victors huzzahed. The skirmishers were promptly deployed; and the rebels, who tried to mislead and draw their fire by displaying blankets and butternut clothing, dodged from tree to tree until the original line was again established.

A professional skulker, who had been unable to elude the vigilance of his officers when this assault was made, threw

---

11 The battle line, under Col. Nelson Miles, 64th New York, actually contained elements of four Second corps regiments and the 11th Mass., which "was ordered there by General Carr for the purpose of feeling the enemy with their sharpshooters. ... we were constantly skirmishing with the enemy during the day, and at about 3 p.m. the enemy commenced massing his troops in two columns, one on each side of the road, flanked by a line of battle about 800 yards in front, in the woods. Their orders could be distinctly heard. They soon advanced with a tremendous yell, and were met with a sure and deadly fire of our simple line. A very sharp engagement continued for an hour, when the enemy fell back in disorder. Their charge was impetuous and determined, advancing to within 20 yards of my abattis, but were hurled back with fearful loss, and made no further demonstrations." [OR XXV/1, 323-324]

12 Blaisdell. [Familiar Quotations]

himself upon the ground, groaning and whining, "Oh! I'm so sick! — oh! I'm so sick!" — but his disease vanished when the foe fled. A man who was loading his musket threw away a cartridge, with a fearful oath about government contractors; and I noticed that the paper was filled with fine grains of dry earth instead of gunpowder. In the thickest of the firing, an officer seized an excited soldier who discharged his piece with trembling hands near the ears[13] and endangered the lives of his comrades, and kicked him into the centre of the road. Trade prospered throughout the day, and the United-States sharpshooters were constantly exchanging their dark-green caps for the regulation-hats which were worn by the regiment. The captain of one of the companies of skirmishers was posted near a brook at the base of a slight ascent upon which the enemy was massed, and there was a scattering fire of bullets which cautioned all to "lie down." While he was rectifying the alignment he perceived with amazement one of his men, who sat astride a log, and washed his hands and face, and then cleansed the towel with a piece of soap which he carried. One sharpshooter shielded himself behind a blanket, and another concealed himself behind an empty cracker-box, the sides of which were half an inch in thickness, exposed his person as little as possible, and felt as secure as the ostrich with his head buried in the sand.[14]

---

13 It seems Blake is speaking of a rear-rank soldier's panicky firing between the heads of the two front-rankers immediately ahead.

14 The historian of Berdan's sharpshooters says of May 2: "An aide informed Lieut. Thorp that: 'Your orders are to advance out on this plank road in the form of a letter V, the point in front, drive the rebel skirmishers in, find their main force and report back.' This was done with E on the left side of the road and K on the right. On this venture, some sharp duels took place. The infantry supports, keeping too close up, suffered considerable, while the active, dodging Sharpshooters wonderfully escaped." [Stevens, 247-248]

At five, P.M., the sentinel in the top of a high tree reported that the rebels were forming their lines of battle upon the left, near the point at which there were three companies, which were supported by small details from several regiments in Gen. Hancock's division of the second corps. Another force menaced those upon the right at the same time. The reserve of the first body of troops held a rifle-pit which had been built across the plank road; and the two lines of battle formed the base and perpendicular of a right-angled triangle, so that the bullets fired by the enemy at one detachment would enfilade the other: but a few large oaks and the formation of the ground afforded a slight protection. The ominous silence of the sharpshooters in front was a sure indication that the main force was approaching; and a rebel officer upon the left brought every man into his place in the ranks by exclaiming to his command, "Forward, double-quick, march!" "Guide left!" The hideous yells once more disclosed their position in the dark woods; but the volleys of buck and ball, and the recollection of the previous repulse, quickly hushed their outcries, and they were again vanquished. The conflict upon the left still continued; and the defeated soldiers began to re-enforce the troops that were striving by desperate efforts to pierce the line, until a company swept the road with its fire, and checked the movement; and only one or two rebels, at intervals, leaped across the deadly chasm. A demand for ammunition was now heard, — the most fearful cry of distress in a battle, — and every man upon the right contributed a few cartridges, which were carried to the scene of action in the hats of the donors. The forty rounds which [normally] fill the magazines are sufficient for any combat, unless the troops are protected by earthworks or a natural barrier; and the extra cartridges, which must be placed in the pockets and knapsacks, are seldom used.[15]

Two companies, including the one to which the author was attached; double-quicked to the rifle-pit; while those who were in the road, and not engaged, loudly cheered to deter the foe: and the men fought behind breastworks for the first time, although they had performed months of labor upon fortifications which were never assaulted. It was after sunset; but the flashes of the rifles in the darkness were the targets at which the guns were fired, until the enemy retired at nine, P.M., and the din of musketry was succeeded by the groans of the wounded. Fresh troops had been thrown upon this point: the flying balls drummed a queer discord when they struck the logs and bank, but the defenders entertained no thought of retreating; and, when the last bullet had been discharged, the bayonet was fixed; and it was a common remark, "When they come near enough, I will use this." The song of the whippoorwills increased the gloom that pervaded the forest; and the pickets carefully listened to them, because the hostile skirmishers might signal to each other by imitating the mournful notes. The rebels gave a yell as soon as they were beyond the range of Union bullets, and repeated it in tones which grew more distinct when they had retreated a great distance and considered themselves safe. The abattis upon the extreme left was set on fire in this prolonged struggle; and a gallant sergeant — who fell at Gettysburg — sprang over the work, and averted the most serious results by pouring water from the canteens of his comrades until the flames were extinguished.[16] The regiment

---

15 In addition to carrying eight days of rations, each man carried additional ammunition. Blake seems to be questioning the necessity for this, however, the meaning of his reference to being behind earthworks is unclear as it relates to ammo expenditure.

16 Blake notes that the fire started on the extreme left, which would have been the normal position of Company B in the battle line. Blake's own

was relieved at midnight, and returned to the bivouac, where the men grasped the haversacks which had remained there since the morning, and eagerly devoured the simple rations with the sauce of hunger,[17] which the lack of food for eighteen hours, and especially the excitement and labor of the day, had rendered intense. Gen. Hancock, who belonged to a different corps, made an honorable mention in his report of their behavior, with which he was perfectly satisfied; and addressed a special letter of thanks to the colonel and the command for their valuable services.[18]

While this small force had thus valiantly performed its duty, and repelled the successive assaults of superior numbers, the musketry resounded two miles to the right; and a spectacle of shameful cowardice was witnessed, which can be rarely paralleled in the history of civilized warfare. The eleventh corps, which was the largest, and held the post of honor upon the right of the army,[19] was stationed behind strong earthworks, which, according to military treatises, rendered one soldier who defends equal to at least three that attack.[20] The yells of Jackson's advancing troops broke these powerful lines a few minutes after

---

Company K would have been third from the left. The 11th Massachusetts lost five sergeants at Gettysburg, including two from Company B (Richard Clink, Miles Short) and one from Company K (Martin Stone). [MSSM/1, 814]

17 The "sauce of hunger" metaphor is associated with *Don Quixote* (1605) by Miguel de Cervantes.

18 Blaisdell's report noted that Hancock complimented the 11th Massachusetts for their efforts of May 2 which cost them 3 killed and 9 wounded. Blaisdell stated that it was impossible to keep Berdan's sharpshooters in the front, so that he was obliged to send forward some of his own regiment, armed with smoothbore muskets, a wholly unsuitable weapon for sniping. [OR XXV/1, 452]

19 XI Corps actually was the smallest in the army at "barely 4,300 men." [Sears 1996, 238] Hooker placed it on the extreme right because he expected the battle to be fought on the left, and I Corps was moving to further support the right. [Ibid. 226, 238, 237]

four, P.M. [May 2]; and the German regiments that composed the main portion of the corps basely fled without receiving a volley, and rushed pell-mell by thousands upon the road to the ford, wholly demoralized by fear, and made no attempt to re-form their ranks. Rebel prisoners who were captured upon the following day assured me that their brigades reached the works without any opposition; and the commanders were convinced that there must be some artifice, because no one could conceive any excuse for the sudden evacuation of rifle-pits of such a formidable strength. The officers of other corps made themselves speechless by striving to rally the "flying Dutchman,"[21] who was no longer an illusion, but a despicable reality; and the cavalry with their sabres, generals and staffs with revolvers, and artillerists with whips and rammers, vainly attempted to stop the disgraceful flight, which was finally checked by the Rappahannock. "Var ish de pontoons?" "Der wash too many mens for us;" "I ish going to mine company," they continually exclaimed. A squad of the fugitives entered the regiment, and repeated the skulker's story, that their commands were "all cut to pieces;" "We are all that are left;" and, "They fought until their ammunition was gone." The colonel [Blaisdell], observing that their muskets were unstained, opened their magazines, and, finding that they had not used a solitary cartridge, denounced their conduct in the strongest language, and remarked, "I should detail some of my good men to shoot you; but they have no ammunition to waste upon your

---

20 Actually, Hooker had warned Howard on May 2 that the right flank of XI corps was poorly arranged: thinly spread and with hardly any entrenchments. Howard replied that he would remedy the situation, but it seems that nothing substantial occurred. Moreover, most of XI corps faced south, not west, the direction of the Confederate attack. For a discussion of Howard's XI corps positioning, see [Sears 1996, 238-247].

21 Legendary ghost ship that disappeared off the Cape of Good Hope when South Africa was a Dutch colony.

worthless carcasses." A cavalry-man halted one of the foreign generals who was dashing to the rear; who indignantly asked, "What for you stop me; you a private, and me a general?" — "I don't care who you are: I should stop you if you had been the devil," the soldier defiantly replied. "That d—d eleventh corps stole my voice," whispered a field-officer upon Gen. Hooker's staff on the next morning.

Gen. Howard, who commanded them, performed his duty, and was cheered by the troops of other corps; but he was oppressed by the feeling that his reputation had been ruined, and he sadly spoke: "Don't cheer for me; I don't deserve your cheers." The poet has truly said, — *What can ennoble... cowards? Alas! not all the blood of all the Howards.*[22]

The Germans sought to escape the censure which the whole army justly bestowed upon them by tearing the badges from their caps, — for the crescent[23] was recognized as the insignia of a poltroon, — and giving the number of one of the reliable corps if they were questioned about the command to which they belonged. The few brave American regiments [in XI corps] shed tears of mortification, and earnestly entreated that they might be transferred to brigades which were composed of their countrymen. Thus a splendid position that had been gained was lost; a large body of troops showed that they could not be relied upon if placed in any post of danger; and the army narrowly avoided the destruction which had been skillfully planned for the enemy.[24]

---

22 *An Essay on Man*, Epistle IV (Pope).

23 The badge of XI corps.

24 Blake's harsh opinion of the predominantly German XI corps was shared throughout the rank and file in the army. It seems they held Howard blameless, while the German troops were made the scapegoats. Long after the war Blake wrote: "The subject has been discussed with bitterness, but after the review of documents published years afterwards, I am satisfied

The division formed its line in the woods upon the plank road, and checked the advance of the foe; and the soldiers marched by the Chancellor House during the night, noticed with pride that the headquarters of their old commander bore the marks of the conflict, and rejoined the brigade which was posted upon the right of the broad avenue.[25] The first line of battle built a slight earthwork with a few spades which had been thrown away by the pioneers of the eleventh corps. The skirmishers began to exchange shots at daybreak upon May 3; and a bullet penetrated the head of a lieutenant who was asleep in the adjoining company, and he never moved.[26] There was a ceaseless roll of musketry: at half-past five, A.M., the batteries emitted destructive charges of canister, and most of the men in the ranks of the support crouched upon the ground while the balls passed over them. For two hours the hordes of Jackson, encouraged by their easy victory upon May 2, screamed like fiends, assailed the troops that defended the plank road, and succeeded in turning their left, and compelling them to retire through the forest, and re-form their shattered lines.[27] There was no running: the soldiers

---

the chief cause of the rout was the inexcusable negligence of officers of high rank in the Corps." [Familiar Quotations]

25 Blaisdell reported that on the evening of May 2 the brigade was on the right of the road, facing west, formed in two lines: 1st Massachusetts and 26th Pennsylvania formed the front line, with Revere's Brigade [Excelsior], and the rest of the 1st Brigade (11th and 16th Massachusetts, 11th New Jersey) forming the second line. [OR XXV/1, 452] Blake wrote that he "slept that night near the plank road to Chancellorsville, about one hundred yards from the spot where Jackson fell. I was awakened by the reports of rifles fired by the pickets of both armies, and this military genius of the Southern Confederacy finished his course." [Familiar Quotations]

26 Two lieutenants of the 11th were killed on May 3: John S. Harris, Company C, and John Munn, Company F. [MSSM/1, 783, 812]

27 Gen. W. Dorsey Pender, who attacked the line closest to the road on the left, where the 11th Massachusetts was in the second line, recorded the action from the Confederate perspective: "My line had not advanced more

fell back slowly, company after company, and wished for some directing mind to select a new position. Unfortunately the national cause had lost Gen. Berry, the brave commander of the division;[28] the ranking brigadier, Gen. Mott, was wounded;[29] another brigadier was an arrant coward;[30] and the largest part of nine regiments were marched three miles to the rear by one of the generals without any orders.[31] The regiments of the brigade, under the supervision of their field and line officers, rallied in the open field near the Chancellor House, which was the focus upon which Lee concentrated his batteries, until the shells ignited it;[32] and the flames consumed some of the wounded who were helpless; and three women, that remained in the cellar for safety, barely escaped the ruins. The brigade was aligned upon the road to the United-States ford at nine, A.M. [May 3], and the men recovered their knapsacks in the midst of a heavy cannonading which still continued. No symptoms of fear were manifested, although the artillery was planted upon the left, in the rear and

---

than 150 yards before the firing became very heavy, but my men continued to advance, and soon it became apparent that the enemy were posted behind a breastwork of logs and brush. This we carried without hesitating. Beyond the breastworks the resistance again became very obstinate, as if we had come in contact with a fresh line, and this, in its turn, was driven back after a short contest." [OR XXV/1, 935]

Col. Blaisdell wrote: "the enemy succeeding in reaching the left flank of our line, our first line was obliged to fall back. Then the second line took up the battle, and struggled long and hard to maintain their position, but were obliged to fall back." [Ibid. 452]

Lt. Col. Tripp of the 11th Massachusetts stated: "At about 6:00 A.M., May 3, the enemy advanced in great force, and succeeded in breaking the first line, which passed entirely over this regiment, which waited until all had passed to the rear, and then commenced firing, ... The regiment offered a desperate resistance, and only fell back when no hope remained to make a successful effort to hold the ground." [Ibid. 455]

28 Maj. Gen. Hiram Berry (2/III), was killed about 7:00 A.M. by a sharpshooter. Command of the division devolved on Brig. Gen. Joseph B. Carr, and Blaisdell took over 1/2/III brigade for the remainder of the battle.

29 Brig. Gen. Gershom Mott led the Jersey Brigade (3/2/III).

the front, from which point most of the shells were hurled; and the force was threatened with capture. The infantry firing ceased at ten, A.M.; and the rebels that had suffered a large loss and achieved a partial success fortified their new position, which was parallel with the plank road, and posted their left in the woods upon the same ground which the division had been forced to leave. The corps was massed at eleven, A.M., near the White House,[33] which was a mile from Chancellorsville; and the remaining hours of the day were frequently disturbed by the picket-firing and an occasional shelling.

The scenes that are always witnessed upon the field of carnage were increased by the fires in the extensive forest, which burned the wounded of both armies, and tortured the just and the unjust. Some soldiers of the regiment, who felt the glow of the raging flames that every second came nearer, and knowing that their comrades were fighting in another part of the line, and the ambulance corps was miles in the rear, calculated the number of minutes which they expected to survive. A small brook of stagnant water, which divided the company into platoons, in the

---

30 Blake's assessment of his own brigade commander, Gen. Joseph Carr.

31 Due to the command confusion following Berry's death, Joseph Revere had marched Excelsior Brigade (2/2/III), and part of the Hooker Brigade (1/2/III), to the rear. Revere was instantly removed from command by an enraged General Sickles, who had raised the Excelsiors himself in New York City. It was shortly after this incident, about 9:00 A.M., that Hooker was knocked insensible when Chancellor House was hit by an artillery shell.

32 Blaisdell described the new line: "The remains of the brigade which could be found were collected about half a mile from our front line and reformed, and were again ready for the conflict. … Our brigade was now formed in column of regiments … about half a mile from the Chancellor house. Here the enemy were checked and made no advance. We remained in this position the night of the 3rd." [OR XXV/1, 432]

33 This house was generally know as the Bullock House (or Chandler's) and was Hooker's headquarters after he was withdrawn from the ruined

morning became a river of life, over which the fire hissed, and vainly darted its deadly tongues to strike the helpless. A rebel and a member of the brigade rested together near an oak, and mutually assisted each other to fight this terrible enemy; and joyfully clasped their scorched and aching hands in friendship, when it was quelled. Colors were captured, and hundreds of the foe threw down their arms, and retreated with the Union forces; and happy squads without any guard were walking upon the road, and inquiring the way to the rear. Three batteries lost most of their horses and a large proportion of their men by the concentration of Lee's artillery, and the bullets of the sharpshooters, who were specially instructed to pick off the animals before they shot the gunners. Several pieces, including one without wheels, which had been demolished, were drawn from the field by details from the infantry. Some of those who were slightly injured returned to their commands after their wounds had been dressed, and fought again. One cannon-ball killed a cavalry-man and his horse: and a shell tore the clothing from an aide, but inflicted no personal hurt; and he returned, after a brief absence, to search for his *porte-monnaie* [coin purse], which he carried in the pocket that had been so suddenly wrested from him.

The corps-color was always waving in front; and Gen. Sickles, smoking a cigar, stood a few feet from the regiment in the road up which the troops had marched from the Chancellor House; and aides and orderlies were riding to and fro, one of whom reported that his steed had been killed. "Captain, the Government will furnish you with another horse," he complacently replied. A rebel officer of high rank,[34] who had

Chancellor House.

34 Not identified. The highest-ranked Confederate officers captured were four regimental colonels, but all on May 3 at Salem Church.

been captured, stopped near the general, and sought to open a conversation, with the following result: —

"General, I have met you in New York."

"Move forward that battery."

"General, I have seen you before."

"The brigade must advance to the woods."

"General, don't you remember" —

"Go to the rear, sir: my troops are now in position."

There were few, if any, stretcher-bearers at the front, and wounded men that had lost a leg or an arm dragged themselves to the field-hospital; and the surgeons of some regiments which had not been engaged in the battle sat upon a log in idleness, and refused, with a great display of dignity, to assist the suffering who were brought to them, because they did not belong to their commands. This shameful conduct, which I often witnessed, exasperated the officers and soldiers; and they compelled the surgeons to discharge their duty in a number of cases by threatening to shoot them. The heat was very severe: and many cannoneers divested themselves of their uniforms while they were working; and a number of skirmishers, who were posted in the open field, and obliged to lie low without any shelter, were sometimes afflicted by sunstroke. "I will win a star or a coffin in this battle," remarked a colonel as he was riding to the scene of conflict, in which a bullet checked his noble military aspirations.[35] "To take a soldier without ambition is to pull off his spurs; I have got my leave of absence now," gladly said an officer, whose aspirations had always been refused at headquarters, when he left the regiment to go to the hospital. The appearance of a rabbit causes an excitement and a chase upon all

---

35 Writing after the war Blake said that it was William O. Stevens, 72nd New York, who made this comment, "and death won the victory." [Familiar Quotations]

occasions, and one ran in front of the line as the action commenced; and the birds were flying wildly among the trees, as if they anticipated a storm; and a soldier shouted, "Stop him, stop him! I could make a good meal if I had him." — "This is English neutrality," an intelligent metal-moulder remarked in examining the fragment of a shell, and explaining the process of its manufacture to the company;[36] while the rebel batteries every minute added some specimens to his collection. The officials in Richmond published at this time an order, directing that the clothing should be taken from the bodies of their dead, and issued to the living. They always stripped the dead and the dying upon every field; and I noticed that one man who had been stunned, and afterwards effected his escape, wore merely a shirt and hat when he entered the lines. The regiment changed its position nine times during the day, in compliance with orders: rifle-pits were erected upon the last line of defence, and no fires were allowed at night.

Several volleys from the pickets ushered in May 4; but the brigade made no movement, and no advance took place. An officer who was going the rounds in the night was surprised to find one of his most faithful men who returned no answer to his inquiries; and, supposing that he had been overcome by fatigue, and fallen asleep, grasped his hands to awaken him: but they were cold with death. The soldier, killed upon his post of duty, rested in the extreme front, with his musket by his side, and face towards the enemies of his country. Gen. Whipple, the able commander of the third division of the corps, was mortally wounded by a sharpshooter who was one-third of a mile from

---

36 There was only one man on the roster of Company K whose occupation was listed as "moulder" – Pvt. Joseph A. Morrill of Manchester, New Hampshire. [MSSM/1, 812] Morrill apparently identified manufacturer's marks on the shell fragments. The British were officially neutral, but arms makers privately sold to Confederate blockade runners.

him; and a priest administered the last rites of the Roman-Catholic Church upon the spot where he fell, in the presence of his weeping staff and soldiers, by whom he was greatly beloved. A brigade made a reconnoissance in the forest at one, P.M., and captured forty sharpshooters who were perched upon the limbs of lofty oaks, and could not descend and escape before this force advanced.[37]

Whenever the picket-firing became active, shelter-tents were removed from the stacks; knapsacks were "packed up;" boiling coffee was swallowed, or thrown away; and the men stood in line, and were ready to obey the order to march before their officers had commanded them to "fall in." Groups of soldiers were writing letters, sleeping, or playing cards; and as one gambler said, "I will bet—" an exploding shell interrupted his sentence; and he added, "a quarter better." The rebels ascertained the location of the trains upon the north bank of the Rappahannock; opened a battery upon them; and a squad of three hundred prisoners uttered a yell of joy when they saw a cannon-ball enter a large tent which was crowded with the dying and disabled. The direction of the firing was changed, and caused utter dismay when some of the [prisoners] were killed by the missiles that were hurled by their comrades in the army of Lee. The pioneers cut roads through the woods for the passage of the troops and artillery, and constructed booths of boughs for the field-hospitals, from which the ambulances were hourly

---

37 Sickles reported that the enemy sharpshooters were soon "driven handsomely by Berdan," but not before General Whipple's death. Hiram Berdan, a brigade commander in Whipple's 3/III division, does not mention the treed sharpshooters in his report, but does record his unhappiness with his support: "On Monday evening, the Eleventh New Jersey, which was acting as our support, was alarmed by firing on our right, and it opened fire upon my Second Regiment, which was deployed in its front, wounding 5 of my men." [OR XXV/1, 392-393]

conveying the inmates across the river. The army-wagons were parked upon the safe bank, and did not encumber the movements of the corps; and the packed mules, which were generally used for the first time, transported the ammunition, while the cooks brought fresh beef to the front by suspending a quarter to a rail.

A dense fog obscured the river upon the morning of May 5, and clouds which discharged a few drops of rain overcast the sky. The skirmishers were hotly engaged at certain points, while the main body was quiet; and the brigade retired at 10.40, A.M., two hundred yards to the rear of its position, and quickly formed breastworks by cutting the oaks down and throwing the earth up. Two sick Germans in the company (they did not belong to the eleventh corps), who were excused by the surgeons from performing any military services, and could not carry their knapsacks or rations for more than two days, or keep with the ranks of the regiment upon an ordinary march, participated in all the fighting; and no command could produce nobler soldiers. A dismal rain drenched the men in the night, and swelled the waters of the Rappahannock so suddenly that the pontoons were endangered, and the communications of the army were seriously menaced; and men were placed in the boats to hold the anchors, and one bridge was taken up to lengthen the others which did not connect with the banks after the freshet. The brigade, which was nearly exhausted by standing in line of battle,[38] evacuated the bivouac in the woods at three, A.M., upon [May] 6th; waded through the mud of the road, unmolested by the enemy; and recrossed the pontoons at the United-States ford at daybreak. The Army of the Potomac, baffled, weary, and footsore, marched in the midst of the storm, which never ceased during the day, and

---

38 Overall 1/2/III brigade suffered 504 casualties: 52 killed, 387 wounded, 65 missing. [B&L/3, 235] The 11th Massachusetts lost 8 killed, 65 wounded, and 3 missing. [OR XXV/1, 178]

enlarged pools into lakes, while ruts became streamlets, and sought rest in the old camps which they had so often quit with high hopes that were not realized.

Chapter XI

The March to Gettysburg

THE ORDINARY labors of the camp were resumed; the old lines were again guarded; and one regiment near the brigade was drilling upon the plain the day after it returned to quarters. As a sanitary precaution, the barracks used in the winter were levelled; the tents were pitched upon ground that had not been occupied; and the blankets of the soldiers were spread upon a network of branches which was eighteen inches above the surface of the earth. The camps, in June, presented a gala appearance; the streets between the companies were adorned with arches and festoons of evergreen; and short pines and savins which had been transplanted from the forest diffused their genial shade. Although the [Third] corps had nobly performed its duty at Chancellorsville, none of its brave members were allowed to receive furloughs or leaves of absence, because an inspector-general, one of those contemptible staff-officers that skulk to the rear in a battle, and display feats of horsemanship to the cooks, teamsters, pioneers, and other non-combatants, falsely reported that it was demoralized.[1] The only fact upon which this unjust action was based was the neglect of many soldiers to salute these gayly dressed cowards whom they despised. Desertions from the enemy were frequent; and an Alabamian swam across the river upon June 6, and brought a gold watch which his captain lent him to regulate the reliefs, with strict orders to be vigilant and

---

1   As of July 1862 an Inspector General was created for each corps, and later at the division and brigade level. They were responsible for examining the condition of the men, livestock, and equipment, and even kept tabs on sutlers and the quality of merchandise sold to the troops. Under Hooker, regiments were graded by the I-G's and a poor performance would result in reduction of leaves and furloughs. [Lord, 113-114]

keep a good watch; an injunction that he followed in letter, if not in spirit.

The terms of service of many regiments which had enlisted for nine months or two years daily expired; and the army was constantly diminished,[2] while Lee concentrated from all portions of the South every available regiment, and prepared for a bold invasion of the North with the most powerful force that the rebels ever organized.[3] Every object south of the Rappahannock was scanned by many eyes: the troops were ordered to keep constantly on hand rations for three days; and when two divisions of the foe marched upon June 3 to Culpeper Court House, the movement was discovered; and at 3.10, A.M., on the following morning, the brigade was notified to "pack up," and be ready to start at a second's notice. The pontoon-bridge was constructed on the 6th near Deep Run, upon the left of Fredericksburg: the soldiers crossed at this point for the third time; and the enemy that had been invisible for a few days issued from the woods, appeared in front, and received a severe fire from the artillery which was posted upon the bank. The regiment was on picket, watching with interest the bursting shells, and discussing the probabilities of a conflict.[4]

"There's a pretty grape-vine ready to set out," said one person who noticed the beauty of the scenery. "Take it up, and

2   The Army of the Potomac lost 53 regiments due to expiring enlistments between Chancellorsville and Gettysburg – roughly 20,000 men (more than its Gettysburg loss). The nine-month units, mainly from New Jersey and Pennsylvania, had been raised in 1862 in response to the emergency of Lee's Maryland invasion. The two-year units were formed in some states in 1861 despite the Federal government calling for three-years.

3   Lee brought about 75,000 men to Gettysburg, fewer than his 90,000 at the start of the Seven Days Battles.

4   When it became apparent that Lee was taking his troops north, Hooker wanted to attack south across the Rappahannock toward Richmond. Lincoln and Halleck rejected this approach. [OR XXVII/1, 30-35]

we'll set it out on your grave tomorrow," lightly remarked one of the group.

Division generals and subalterns seldom know more than the rank and file about the intentions of the commander of a large body of men: and the regiment was engaged in making preparations to celebrate the 13th day of June, which was the [two-year] anniversary of its muster into the service; and many of the line officers were playing a game of base-ball,[5] when the adjutant arrived with marching orders; and within half an hour tents were struck, knapsacks packed, and the column in motion at 1½, P.M. The general commanding the brigade[6] pushed forward the troops in the most unmerciful manner, and great joy was manifested when they bivouacked near the Hartwood Church.

Hundreds of blankets and overcoats were left upon the field in the morning of the 12th; and the natives and negroes of both sexes collected them as soon as the march was resumed. Caligula[7] and other monsters of antiquity never displayed a more diabolical spirit than certain generals in the corps, who murdered the unfortunate soldiers that were compelled to obey their orders, by exhausting their strength, and needlessly exposing them to the rays of the sun, which through their cruelty, became as deadly as *Minié* balls. There is not more than one in ten officers of high

---

5   Capt. Mason W. Tyler (37th Massachusetts) noted that "by 1863 baseball was 'all the rage' now in the Army of the Potomac…almost every street having its game." [Kirsch, 37] The *New York Clipper* of June 20, 1863 reported on a game played May 30th between the 11th Massachusetts and 26th Pennsylvania for a $50 prize (won by the 11th). "We have four clubs in our brigade, and there are several more in the division."

6   Brig. Gen. Joseph B. Carr, whose brigade had lost 2nd New Hampshire but gained 12th New Hampshire and 84th Pennsylvania. Replacing Hiram Berry as division commander was Andrew A. Humphreys, who had previously commanded a V corps division of predominantly nine-month regiments whose terms had expired.

7   Roman Emperor from 37 to 41 CE, known for his cruelty.

rank that understands the proper mode of moving divisions; and the fatigue that so often results is caused, not by merely travelling a large number of miles, but by the omission to halt them at regular intervals after marching short distances. Mounted upon their horses, unencumbered by rations or clothing, and usually carrying a small flask and a light sword, it was a pastime for the subordinate generals and their staffs to ride or race from town to town, and issue stringent orders to court-martial the weary men for what they termed straggling. The division marched from 5.20, A.M., until 9.20, P.M., [June 12th] upon one of the warmest days of the month, and was always designedly halted in the open fields, while a general and his staff enjoyed the comfort of the extensive forests in the vicinity, and, with the hearts of demons, laughed and uttered jokes about the soldiers who were dying or writhing in the agonies of sun-stroke. Some surgeons and regimental commanders remonstrated against this inhuman conduct, and told a general that it was killing the men; but he sneeringly remarked; "If I can stand it, the men can;" or, "The sun will dry their shirts, if they are wet with perspiration." I speak in emphatic language, for I recall the forms of prostrate heroes who had escaped the bullet, the shell, and the "fiery darts"[8] of the foe in scenes of combat, and fell fainting by the roadside, to die, or linger in pain from which they never recovered. Thousands of throats were sometimes screaming, "Halt, halt!" and there were universal cries, "Kick him out of the house!" "I hope the rebels will kill him!" "Shoot the scoundrel!" interlarded with the most profane oaths ever uttered, when the command passed by the mansion selected for headquarters. Some men seized the general's servant, who had walked a long distance to procure cool water, and spitefully confiscated the property.

---

8 *Paradise Lost*, Book VI (Milton).

Less than one-third of the division bivouacked at Beverly Ford;[9] and the stream of soldiers that had been forced by the foregoing causes to leave the ranks was continually flowing into the regiment during the night.

The cavalry had crossed the Rappahannock at Kelly's Ford and this point on the 9th, and fought a successful battle, which completely frustrated the plans of Stuart, who had assembled his force near Brandy Station for the purpose of making a formidable raid. The rebel pickets rested on their horses in the rifle-pits upon the opposite bank; and the presence of the enemy intimidated the general who had been so eager to place his troops in the open fields upon the march. The camp-fires were prohibited or kept low, and strong guards were posted to prevent them from leaving the woods in which they were concealed. The river was only one hundred feet in width at the ford, which was protected by a rude breastwork of rails. Wounded horses were limping about on the ground in the vicinity; the carcasses of dead animals that had "fought like men" were scattered in every direction; and I saw one floating in the stream, that was fully equipped, and still bore the rations, blankets, and overcoats of its absent rider. Redoubts for four guns were erected in the night of the 13th by details from the brigade. The newspapers contained, at this time, accounts of the operations of Gen. Grant, which resulted in the capture [siege] of Vicksburg, and described the scanty wardrobe with which he was furnished upon the campaign; and the contrast between this simplicity, and the immense quantity of personal baggage which the general commanding a brigade carried in the wagons, was as striking as the difference in courage and military ability. Whenever a halt was ordered at the end of a march, a

---

9 The distance from Hartwood to Beverly Ford is about 17 miles. 2/III
   division was to guard the crossings of the Rappahannock from Wheatley
   Ford near Kellysville to Beverly Ford. [OR XXVII/2, 530]

score of servile pioneers pitched his capacious tent upon the most pleasant spot of ground, and placed in it a carpet, camp-chairs, tables, and an iron bedstead, so that he was probably more comfortable than he would have been at home.

The brigade was relieved by the cavalry after sunset on the 14th, and marched throughout the night until 7.10, A.M., of the 15th, to Catlett's Station, where it remained an hour and a half; and moved again at 8.40, A.M.,[10] until it bivouacked near Manassas Junction after midnight.[11] The division was in the rear of a train of wagons which constantly obstructed the road and interrupted the passage of troops, so that a general was obliged to halt when he wished to advance. There was no enemy in front; and the quartermasters, commissaries, and other staff-officers, who are never seen near the general in a battle, accompanied him upon the march; and the number of his victims was increased by the same infernal conduct that has been already noticed. The soldiers were forced to halt in the fields, without any shelter from the sun; and those who were overcome by the severe fatigue which always follows a movement in the night, and fell asleep, awoke to suffer from an intense faintness and pain, which disabled them for years.[12] There was sometimes a solitary oak or pine that stood upon the plain like a rock in the desert; and the limits of the shadow upon the ground enclosed a small squad that crowded together to enjoy the protection of the enlivening

---

10 Humphreys and Carr, in their reports, indicate that the stop was longer, stating that the march resumed at 2:00 and 3:00 P.M., respectively.

11 Direct distance from Beverly Ford to Manassas Junction is about 25 miles. Chaplain Cudworth said that the marches in mid-June 1863 were "at a rapid pace, through clouds of dust in some place so dense, that vision was impossible beyond a few yards; and a large number of men fell out by the wayside, utterly overcome by heat and fatigue." [Cudworth, 383-384]

12 Both Carr and Humphreys noted the severity of the march and that many men suffered sunstroke. [OR XXVII/1, 530] Ironically, Humphreys had been disabled by a heat stroke early in his career.

foliage; while others slept in the dust of the road, beneath the wagons. Many a soldier uttered the wish that he might be permitted to serve out the remaining months of his term of service in some prison, or be changed into the general's horse, when he beheld that officer with his staff and their steeds reposing in the vast forest from which they were excluded. The region was unusually dry; and the men, urged by necessity, slackened to a slight extent their thirst by extracting the moisture from the lumps of damp earth. Thousands were exhausted, and sought the woods to recover their strength; but a general issued orders to the provost-guard to set the dry leaves on fire, and thus drive the members of his force into the cleared tracts of land, and clouds of smoke rolled upwards in the rear of the column. The combined heat of the rays of the sun and the burning forest was unendurable; the breath seemed to be a flame; and less than one-fourth of the division rested at the bivouac.[13]

No movement was made upon the 16th [of June]; and the troops fixed bayonet, stuck them into the ground, and put up their tents by fastening pieces of shelter to the muskets. In accordance with orders from the headquarters of the army, the bands and field music performed the principal portion of the day to mislead the scouts of the enemy, so that they could not ascertain the point at which the largest force was concentrated. In addition to the regular rations which were issued to the soldiers, they lived upon the resources of the country, and devoured many highly seasoned dishes of frogs and box-turtles. The lines moved a short distance

---

13 "The daily marches were usually long, and made at an unusually rapid pace; so that the roads were lined with stragglers representing almost every regiment, some of whom had been sunstruck, and were completely broken down. To add to the discomforts ordinarily experienced, the woods and fields had been set on fire, intentionally or otherwise, which filled the atmosphere with smoke and cinders, compelling the soldiers to bivouac upon the open plains." [Cudworth, 385]

to the right on the 17th and 18th; and a general, to gratify his savage disposition, placed his staff at the dry crossings near Blackburn's Ford [on Bull Run Creek] to force the brigade to wade through the stream; but the use of the bayonet and musket upon the horses defeated this design, and the enraged officer remarked, "The men acted like sheep."[14] The companies were now commanded to attend the roll-calls eight times daily; the names of absentees were reported for punishment; and, while the regiment was pitching its tents upon an established line, marching orders were received at two, P.M., and the column bivouacked at the end of eight hours near Gum Spring.[15] A general who was alarmed by the dense clouds that darkened the earth deserted his troops that were moving upon the road, galloped to the village with his staff, and selected the largest edifice for his headquarters, while his command under other officers was posted on an important picket duty in the midst of one of the most copious rains which ever fell in the South. A field-officer on the corps-staff,[16] who frequently said, "Boy and man, I have been in the regular service twenty-five years," and acquired the sobriquet of "crazy" on account of his lack of common sense, flourished a revolver to drive the men from wells on the march, and acted the part of a useless and harmless non-combatant in the time of action.

Gen. Braddock bivouacked his little army at this point [Gum Springs] while he was marching upon his ill-starred expedition,[17]

---

14 It is not clear what Blake means by this — apparently the men chased the horses from the crossing by brandishing their weapons.

15 Now called Arcola; the distance from Manassas to Gum Springs is 18 miles. 2/III division paused at Gum Springs for six days.

16 Unknown. Possibly Lt. Col. Julius Hayden, who was "more noted for frouth and foam than for common sence." [Pearcy, 24]

17 In June 1755, during the French and Indian War, British General Edward Braddock set off from Virginia to capture Fort Duquesne (near modern-day

and gave the name to the hamlet, the old inhabitants said, from the ever-flowing spring, which was celebrated for its purity, and a gum-tree that was still standing, although many of the branches were rotten, and showed marks of its advanced age. Guards were placed over it to prevent the soldiers from enjoying the priceless liquid without a permit in writing from the general. The drivers, who daily renewed the water in the kegs which are attached to the ambulances to convey a fresh supply for the weary, the sick, or the wounded, were pushed aside at the point of the bayonet, and directed to go to a turbulent run in which horses and mules were standing or walking, and the men were washing their persons and clothes. Twenty members of this [ambulance] corps were performing this labor at the streamlet; and I heard one of them say (with many oaths, as a matter of course), "I hope Gen. _____ will be shot, and live just long enough for me to pour my keg-full of this gravel down his throat!" a righteous sentiment, which was reiterated by the score of tongues.[18]

The division, leaving hundreds of sticks, barrels, and boards which are always seen upon every field in which troops have bivouacked, commenced a march [on June 25th], which, like those I have described, will never be forgotten for its severity. It was a ceaseless tramp for eighteen hours, with only one regular halt, that occurred in consequence of a misunderstanding regarding orders. The Potomac was crossed at Edward's Ferry upon a pontoon-bridge consisting of sixty-four boats, more than a quarter of a mile in length, and travelled upon the narrow towpath to the mouth of the Monocacy, and bivouacked after midnight, twenty-nine miles from the point of departure. The rain

---

Pittsburgh), but was soundly defeated by the French and died four days after the battle.

18 The object of the soldiers' wrath was Humphreys: 2/III division commander. [Familiar Quotations]

in the latter part of the night created new obstacles: men were continually falling from utter exhaustion, and limbs were sometimes broken upon the sharp rocks. A few slipped into the canal; and some who were always "jolly" under all circumstances greeted the ears of their unfortunate comrades, when they rose to the surface, with the consoling words, "It will take a month's wages to pay for that musket." The towpath, which did not exceed twenty feet in its width, had been constructed between the Potomac River on one side, and the Baltimore and Ohio Canal upon the other; and two generals chuckled, and said that there could be no straggling in such a place, and dashed ahead of the column to secure a pleasant repose for the night. The physical power of the soldiers was reduced to such an extent, that when the regiment was halted by its commander,[19] who acted without orders in their absence, only eighteen muskets were stacked in a line that should have had 325 arms: two line officers were present, and the other organizations of the division were scattered in the same manner.[20] A general greeted the appearance of the remnant of his command in the

---

19 Lt. Col. Porter D. Tripp. Blaisdell was on sick leave due to chronic bronchial condition.

20 From Gum Springs to the Monocacy-Potomac confluence is about 30 miles. "For length, severity, and discomfort, [the march] exceeded anything the army ever had been through before… there was no place to rest with any comfort; and therefore the march was kept up, at a quick pace, until one o'clock, A.M. Three hundred and sixty men belonging to the 1st [Massachusetts] left Gum Spring in the morning, but only forty laid down in the rains, seventeen hours after, on the banks of the Monocacy." [Cudworth, 386]

 Pvt. Thomas V. Cooper, 26th Pennsylvania, recalled: "No man who participated in that march can ever forget the driving rain, the slippery and narrow pathway, with water to the right of us, water to the left of us, water above, water below — without opportunity to halt, or rest, or eat or drink, until the late hour found us at our destination."

 Pvt. Walter Morss, 11th Massachusetts, summed up the day's march in his diary: "Very stormy. May I never see another Tow Path."

morning with hearty laughter, and uttered many gibes concerning their weakness, and the condition of those who were without shoes, and bound handkerchiefs and towels around their feet, or wore two or three pairs of socks; and made this remark about the regiment as it filed by him: "What regiment is that? Bring them here, and we will pray for them."

The column steadily advanced upon the remaining days of the month of June, and bivouacked upon the 30th at Bridgeport, [Maryland] after proceeding through Burkettsville, Middletown, Frederick City, and Taneytown.[21] The Potomac, that separated Virginia from this section of Maryland, was the boundary between institutions as conflicting as slavery and freedom, or ignorance and intelligence. The soldiers had witnessed for two years, in the first State, barren lands, a treacherous and benighted race, children in rags and filth, miserable roads, and rude cabins of the "poor whites" and African bondmen, and empty churches; for the bells were cast into cannon, and religion and morality were sacrificed to gain Southern independence. The scenes were changed so suddenly, that it seemed like a delightful vision to behold the schoolhouses; the noble faces of the people; the splendid streets of a civilized age; the cultivated farms and orchards; the cottages ornamented with flowers; and, above all, the smiles and words of welcome from loyal men and women who publicly displayed the American flag, gave refreshing water to the soldiers while they were marching, and refused in many places to accept any compensation for food. A number of mills were in operation upon Sunday to supply the inhabitants with flour, because they had exhausted their store in making bread for the Union army. The brutality of the generals was almost forgotten; and weary feet regained their strength when they

---

21 The five-day march from the Monocacy to Bridgeport (about 37 miles) required a mere 7.5 miles per day.

touched the soil and moved over mountains; while the eye saw, in the magnificent valleys, communities that resembled their homes in New England. The conduct of the troops, with the exception of many non-combatants, the army-thieves, and plunderers of the dead, was unexceptionable; and no profane or improper expressions were heard by any of the citizens. A general[22] placed sentinels upon the houses in many towns, not to protect property, but to obtain for the use of his mess all the luxuries which they contained; while the line-officers and enlisted men were driven from the premises, and not allowed to purchase articles of food. One general in the division [Carr], well known for his cowardice, marched through the populous districts with much ostentation at the head of his brigade, and shouted orders in a pompous tone of authority to attract the notice of the crowd; while the soldiers were saying, "It is perfectly safe to be in front now;" "There won't be any fighting while he leads the brigade," and similar sentences. This officer had taught dancing schools of a low character before the war; and the members of some companies would "call off" the various changes, — "Right and left," "All promenade to the bar," &c., — whenever he rode by them, for the purpose of insulting him.[23] The herds of cattle from which the supply of fresh beef was obtained moved upon the roads with the trains, bearing upon their horns and backs the knapsacks and muskets of the guard, and followed the leading ox, which was conspicuous on account of its size. At other points of the march, several hundred cavalry-men were asleep upon the ground at the feet of the horses, with the reins in their hands; and

---

22 Probably Humphreys — consistent with his policy of guarding the waters at Gum Spring.

23 Ex-dancing masters in the army were also noted by Lt. Col. Theodore Lyman, who pointed out that Carr was an Irish dancing master and Edward Ferrero was of the Italian variety. [Lyman, 180]

I was informed that no one was ever injured [by their horse] in this position. The rebel cavalry committed many outrages in the tracts which they visited; demolished brick ovens, and plundered dwellings, like thieves; but sometimes behaved with moderation; and, in one village, riddled with bullets an innkeeper's sign upon which the American eagle had been painted. The attention of an officer of high rank was called to the large number of deaths and casualties from sunstroke and exhaustion in a certain command; and the generals were compelled to halt a few minutes in every hour; and long distances were thus marched with ease.[24]

The army received the news of the removal of Gen. Hooker, and the appointment of Gen. Meade [June 28], with amazement, and refused to believe the fact until the orders were read; and the opinion was expressed that he had fallen victim to the implacable hatred of Gen. Halleck and the machinations of Pennsylvania politicians.[25] When we consider that the corps were marching to encounter the enemy, and daily expected to fight the decisive battle; that Gen. Meade was unknown to the troops, and had never commanded under a heavy fire a body of infantry exceeding a division; that other officers were superior in rank, capacity, and experience; and that Gen. Hooker had made every movement with consummate ability, — it was an act of the most hazardous character. If the question is viewed from a military point of view, and it is remembered that Gen. Halleck, untaught by the surrender of Harper's Ferry in the previous year,[26] wished

---

24 Possibly referring to Birney, acting III corps commander (ex-Sickles, temporarily absent), who issued a circular on June 17 stating in part that officers leading a column shall "halt sufficiently often to keep it well closed." [OR LI/1, 1060-1061]

25 Blake's unceasing loyalty to Hooker was probably typical of III corps veterans. The battle of Chancellorsville had not diminished their regard.

26 During the Antietam campaign in September 1862, the Harper's Ferry garrison of 12,000 men had been captured when surrounded by Stonewall

to maintain at that post a large garrison and withhold it from the main force, the views of Gen. Hooker are so clearly right, that all must conclude that it was a shallow pretext for the unjust removal of one of the most loyal and gallant soldiers that the country ever produced.[27] The rebels were delighted with the change; and while Lee denies in his report that his cavalry was defeated in any engagement, he admits that the dispositions of the army by Gen. Hooker completely baffled his plans for the capture of Washington, and forced him to fight the battle of Gettysburg. The system of his [Hooker's] mind was such, that the succession in commanders caused no delay in the advance of the different columns.[28]

The division reached Emmettsburg[29] upon July 1, and the beautiful clouds upon the summits of the mountains seemed to be within the grasp during the rain that ensued. At ten, A.M., I heard the report of a cannon which was discharged in the State of Pennsylvania when the first conflict took place between the cavalry and the vanguard of Lee's grand army;[30] and there was a general feeling of relief that the long marches were ended, and the foe, that must be fought at some point, was preparing for the

Jackson's corps.

27 Hooker's stated reason for resigning was Halleck's refusal to abandon Harpers Ferry and assign the garrison to his army.

28 In other words, Hooker's marching orders were maintained after the change of command. Never had the Army of the Potomac marched so far and so fast as in June 1863. The rapid march to Gettysburg was more than triple its rate of march of 1862. This increase from about six miles per day to twenty-plus per day severely upset Confederate expectations, and may account for Lee's aggressiveness at Gettysburg.

29 Bridgeport to Emmitsburg is a march of about 23 miles.

30 Humphreys wrote: "while I was engaged in a careful examination of the ground in front of Emmitsburg, the division was ordered to move up to Gettysburg, 12 miles distant, where an engagement had taken place between the two corps of General Reynolds [I] and Howard [XI] and the enemy." [OR XXVII/1, 530]

most desperate battle of the war. A squad of rebel prisoners passed to the rear, and the usual salutations were exchanged, "How are you, Johnny Reb!" "How are you, blue-belly?" The brigades of the division were posted in Echelon, after manœuvring two hours; sentinels were stationed to guard the lines; and the troops, stimulated by the heavy cannonading upon the right, advanced towards Gettysburg at four, P.M.[31] A negro, who was greatly excited and scarcely able to speak because he knew that the soldiers were marching in the wrong direction, earnestly said to a general, "The road is full of 'em, — heaps of rebels!" but that officer[32] avowed his leading principle to be, "Never believe a nigger;" and the column pushed on. The regimental band played, "Home, sweet Home," when the boundary-line was crossed; and the Twenty-sixth Pennsylvania Volunteers, which formed a part of the brigade, and conferred high honor upon its State by bravery upon many a stubborn-fought field, greeted their native soil with enthusiastic cheers. The infantry was forced to wade through Marsh Creek several hundred yards, and not allowed to pass over the covered bridge; while a general and his staff sat upon their horses, and amused themselves by laughing at those on foot in the stream. A citizen remarked, "If you go on, you will have to fight in the night;" and one of the rebel pickets who was searching for water found himself a prisoner in the hands of the advance guard at ten, P.M. The regiments were at once halted in the road, and ordered not to talk or light matches; while the mounted officers above described, including "Crazy," promptly retired to the rear; and the three long miles which had been uselessly travelled were

---

31 Only the 1st and 2nd Brigades marched to Gettysburg on the evening of July 1. The 3rd Brigade joined the division at 9:00 A.M. July 2.

32 Humphreys again. [Familiar Quotations]

retraced in silence.[33] Willoughby's Run was forded; the vedettes of the cavalry were passed within a short distance of the blazing camp-fires of the enemy; and the division joined the third corps, and bivouacked upon the plains of Gettysburg at half-past two, A.M., on July 2.

---

33 The Confederate searching for water was "captured near Black Horse Tavern, about two hundred yards from Gen. Longstreet's Corps ... There was a sudden halt, orders were whispered, talking, smoking and lighting of matches were forbidden, cups, dippers, pans and articles were tied together to keep silence in the ranks. The generals and their staffs riding proudly at the head looked and rode backward and the miles traveled uselessly were retraced." [Familiar Quotations]

Humphreys was relying on one of his staff officers: "Lieutenant-Colonel Hayden was positive that General Sickles had instructed him to guide the division by way of the Black Horse Tavern ... I found myself in the immediate vicinity of the enemy, who occupied that road in strong force. He was not aware of my presence, and I might have attacked him at daylight ... but I was 3 miles distant from the remainder of the army, and I believed such a course would have been inconsistent with the general plan of operations of the commanding general." [OR XXVII/1, 531] For additional contemporary observations on Humphreys character and style of command, see *Red-Tape and Pigeon-Hole Generals*.

Chapter XII

The Battle of Gettysburg

DURING the night [July 1st and 2nd], the picket-firing did not interrupt the sleep of the soldiers, who were astonished when the morning came to see the Union skirmishers advance and receive volleys from the enemy, that occupied the road over which the division had marched five hours previous. The batteries were pointing in the same direction, and the first movement which the regiment executed was a countermarch, so that it faced the foe, that slowly deployed its columns in line of battle, until the incessant rattling of the rifles of sharp-shooters and those upon the outposts gradually extended from right to left along the vast front.[1] Some regiments were detailed to leave their stacks and equipments and demolish the rail-fences which had been constructed upon the large open field, and would be obstacles to the quick movements of the troops or artillery. The unpleasant mist and the clouds that threatened a storm at daybreak disappeared before noon, and both armies were engaged in the different manœuvers which always precede a battle. The principal portion of the inhabitants deserted their houses with their families, and fled many miles to places of safety; while others rendered good service by acting as guides for the cavalry and the national forces.

The pencil of the artist has portrayed the topography of the scene of conflict and indicated the position of all the troops with

---

1 At 11:00 A.M. Sickles ordered Carr to send a regiment to support the picket line: the 1st Massachusetts was sent and relieved a regiment from Birney's division. Humphreys wrote that shortly after noon he was ordered to form his division in line of battle "near the foot of the westerly slope of [Cemetery] ridge." [OR XXVII/1, 533]

such wonderful accuracy, that no pen can make its history more complete. The corps advanced in a brilliant line half a mile at three o'clock in the afternoon,[2] and the regiment was formed upon the Emmettsburg Road, and partially sheltered by the house and barn of Peter Rogers, upon the crest of the rising ground.[3] The enemy was concealed in the forest, and the main force was unusually quiet until the rebel skirmishers applied the torch to some houses, and the consuming flames and clouds of smoke excited yells of joy. The eye beheld, in every direction, battery and brigade extended from point to point; the moving columns and gay banners; the white marble monuments in the cemetery upon the right, that contrasted strangely with the glistening cannon; the signal-flags that were waving from the craggy summit of Roundtop Mountain upon the left: but there were no tragic pictures of human strife, and it appeared to be a peaceful review. A herd of thirteen or fourteen cows was quietly grazing upon the field; flocks of tame pigeons sat upon the dovecots and sheds; and the lady who lived in the cottage was baking bread; and sold chickens to soldiers in the regiment.[4]

A rebel battery opened an enfilading fire upon the brigade at forty-five minutes past three, P.M., with solid shot, which were discharged from the pieces at a depressed angle, struck the earth,

---

2 Blake's time notations conflict with Carr's report, saying "it was at "4:08 p.m., by order of General Humphreys, I advanced my line 300 yards to the crest of the hill [on the Emmitsburg Road]." [OR XXVII/1, 543] Humphreys' 2/III division had been ordered to Little Round Top by Gen. Meade. In a post-war letter Humphreys wrote that "the division marched by the left flank for a short period of time, then the order was canceled, and the brigades reversed direction and formed on the Emmitsburg Road." [Humphreys Bio, 192-194]

3 This position is the location of the 11th Massachusetts monument: on the east side of Emmitsburg Road at the corner of Sickles Avenue.

4 The lady was Josephine Miller, granddaughter of Mr. Rogers, and the only female member of the III Corps Veterans Association.

bounded into the air, and leaped like a rock skipped upon the surface of the ocean by the powerful arm of a giant. The balls penetrated the building within a few minutes, and one shattered the oven; but the woman was undaunted, and exclaimed, "I will never leave this house," and retreated to the cellar at the request of an officer. Her husband, who had been trembling with fear for hours in his place of refuge, whiningly said that it was strange that they could not fire over his dwelling, and not through it.

The great contest began upon the extreme left, and soon raged with such intensity that the troops were enveloped in the smoke of battle; and it was evident that Lee was exerting every effort to gain the Roundtop Heights,[5] from the summit of which a battery that had been drawn up the abrupt and stony sides with immense difficulty belched forth shell and canister into the corps of Longstreet. The skirmishers in our immediate front[6] reported that the rebels were massing their brigades for an assault upon the position held by the division; and the men, without erecting breastworks, prepared to resist the onset: and every one, knowing the vital importance of the pending struggle, stood firmly upon his foot of ground, which he determined that he would never yield. The batteries and infantry which were posted on the extreme left were steadily driven towards the centre, and were rapidly moving half of a mile in the rear of the division before the yells and bullets of the enemy showed that the long-expected line was advancing. Soldiers who had been forced to leave the ranks upon the exhausting marches continually rejoined their commands, and some without muskets were waiting to seize and

---

5 Up to this time the action had occurred further left, with the Confederate divisions of Hood attacking toward Round Top and Devil's Den, and McLaws through Rose Woods toward the Wheatfield and Peach Orchard. These latter points were defended by Birney's 1/III division.

6 1st Massachusetts performed skirmish duty for Carr's Brigade.

use the arms of those who should be killed or disabled. A snake that rustled through the grass at this exciting moment was promptly despatched by a squad whose minds were not discomposed by the perilous state of affairs. The skirmishers fell back to the main line, which was calmly resting in the road, and holding its fire until the rebels should reach and attempt to climb a rail fence in front. The regular battery, planted upon the left of the regiment,[7] decimated their ranks with terrible charges of canister, that swept the field again and again, and caused a cloud of dust; and all thought that the repulse might be decisive. When the musketry riddled the house, a kitten, mewing piteously, ran from it, jumped upon the shoulders of one of the men, and remained there a few minutes during the fight.

Before the regiment could deliver its volley, the companies about-faced in pursuance of the orders of some stupid general,[8] and executed a right half-wheel under a severe fire, with as much regularity as if they had been upon parade, and thus abandoned the advantages of the strong line of defense in the road. The "stars and bars" [flag] of treason were visible when infantry could not be seen; and the column which had been shattered by the battery appeared in front and began to shoot the gunners, who performed their duty with the utmost fidelity, and retired at last to escape the capture which seemed to be unavoidable.[9] While the rebel standard-bearers waved their colors, the officers beckoned with their drawn swords, the men with their hands exultingly pointed to the divisions that were flying from the left, and sought

---

7   Humphreys had requested additional artillery and he had been sent Lt. John G. Turnbull's batteries, F and K, 3rd United States.

8   Although Blake is probably referring again to Carr, the order for this maneuver originated with Birney who had assumed command of the Third Corps when Sickles was seriously wounded. Birney was attempting to align Carr's brigade with the retreating troops from his division on the left.

9   Turnbull's battery lost 9 killed, 14 wounded, and 44 horses killed.

by their shouts and gestures to encourage the timid and quicken the march of the support, and the soldiers were constantly loading and aiming their rifles at the breasts of the members of the regiment, orders were duly transmitted from a blockhead, termed upon the muster-roll a brigadier-general, not to discharge a musket, because they "would fire upon their own men;" and the enemy was enabled in this way to cut down the ranks, and diminish the effect of the first volley.[10] Candor compels me to admit that the officer from his standpoint, which was far in the rear, could not distinguish one line of battle from the other. The command was disregarded: the foe stood in groups of three or four, and the large number of gaps or intervals which were not closed up revealed the extent of the slaughter; and the survivors, always seeking, like Indians, a hiding-place, entered the road, sought the protection of a slight ridge, and their advance was entirely checked.

A heavy mass of infantry appeared upon the right of the house at this glorious moment,[11] and the new formation of the regiment exposed the line to an enfilading fire which was very destructive. The right of the brigade was not within the supporting distance of the second corps: the left of the division had been forced to fall back, so that the troops were subjected in certain positions to volleys from three distinct points, and the men slowly retreated, foot by foot; while thousands, pierced by

---

10 Blake explained in his post-war writing: "A building in the extreme front, and west of the Emmittsburg Road, was occupied by pickets of the Sixteenth Massachusetts, and an order issued from division headquarters, cautioning all the reserves not to fire on these comrades when they retired." [Familiar Quotations] Carr did place his 1st Massachusetts skirmishers in advance of the 26th Pennsylvania, a move Col. Clark B. Baldwin of the 1st Massachusetts did not understand or appreciate. [Baldwin]

11 The attacking Confederates were two brigades of Anderson's division: Wilcox's Alabamians north of Roger's house, and Lang's Floridians slightly south of the house.

the deadly *Minié* balls, or torn asunder by the explosion of the infernal shell-bullet, fell, and saturated the plain with their blood. As soon as the bullets began to whistle, a general [Carr] said to the orderly who carried the color of his brigade, which he supposed would attract notice and draw the fire of the enemy upon him, "Take away that flag;" "Go to the rear with that flag;" and the person who obeyed this direction remarked in stating it, "Faith, an' I was as willin' to run with it to the rear as he was to have me." The most demoralizing results would have occurred if the troops had been new when this event took place; but they were veterans, and the shameful misconduct of the officers who commanded them did not affect their constancy or firmness. The long distances over which the rebels marched to make their grand charge, and the serious losses which they sustained when they gained the Emmettsburg Road, had reduced their numbers and strength, so that a vigorous attack upon their left flank by the second corps, the concentration of the batteries that were posted upon the interior lines, and the resistance of the troops that rallied, repulsed them at sunset. At this critical time, in obedience to a universal cry among the soldiers, "Charge on them!" "Take our old ground!" the fragment of the brigade, with the colors of five regiments unfurled within the distance of one hundred feet, in the absence of its general, and against the orders of Gen. Humphreys, the division commander, who vainly shouted, "Halt, halt! — stop those men!" pursued the enemy half a mile, captured several hundred prisoners, retook cannon that had been left upon the field, and assisted to achieve a conclusive success. Those who suffered from fatigue in retreating before a victorious foe until they could barely move recovered their strength when the circumstances were reversed, and they gladly ran to overtake the defeated force. Ten thousand of the dead and wounded of both armies were mingled together upon an open space of ground, less than three-quarters of a mile square; and it was sometimes almost impossible to advance without walking upon the form that four hours before had been strong with life, and animated by its high hopes. The disabled Union soldiers and some of the enemy expressed their joy, and uttered many welcomes, when the troops followed the receding lines; and there were cries, "Go in!" "Go in!" "Drive them from the field!" "I don't care for my wound, if we only whip them." The rebels told me that their generals and officers said that there was nothing in their front except a force of militia, which would run away at the first volley; but this falsehood was detected as soon as the fighting commenced.[12] They deceived others, who implored the national troops not to kill them.[13] I observed one wounded youth about sixteen years of age, who was crying, and stated the cause of his grief, that "Gen. Lee always puts the Fifth Florida in the front."[14]

The batteries of the enemy were very active, and furrowed the field with shot and shell which mangled the bodies of the dead and dying; and those who could not move had crawled into little gullies, or protected themselves behind the rocks, which were numerous. When the flying rebels disappeared behind the crest of the elevation near the Emmettsburg Road, members of both armies, who had thrown themselves upon the ground from exhaustion and other causes, and were stretched upon the plain apparently lifeless, rose uninjured in every direction. The enemy had examined the [Union] officers, and sent to the rear as

---

12 Lee was somewhat "in the dark" - not realizing that the entire Army of the Potomac was in Pennsylvania by July 2.

13 Apparently some rebel commanders told their troops that the Yankees would kill their prisoners.

14 It got worse for Lang's Florida Brigade the next day, supporting Pickett's charge. All regiments of the brigade lost their colors at Gettysburg, except for the 5th Florida.

prisoners those on whom they found no wounds. The Union soldiers immediately inspected every man that wore the butternut uniform, and discovered many who were feigning severe bodily injuries by uttering groans and similar devices. Squads seemed to be terror-stricken, and dodged or crouched upon the earth whenever the shells that were fired by their batteries exploded near them. This reverse was so unexpected by Lee, that three pieces of [U.S.] artillery upon the left of the regiment, which had been abandoned when the troops that supported them were forced from their position, had not been removed, and were captured by the troops; and one was retaken, together with twenty rebels, who had pushed it about one hundred yards. The prisoners assisted those [Union soldiers] who were pulling the cannon from the field, and gladly rushed with it to the reserve to escape the storm of shot.[15] The wounded that were not utterly helpless slowly traveled to the hospital; and the ambulance corps, with the exception of a few faithful stretcher-bearers, did not render any aid to the others.

---

15 "Hundreds of Confederates, uninjured and hugging the ground to escape from the shot and shells of their artillery, were compelled to stand up and abandon their guns and ammunition under orders of 'Go to the rear,' 'Hurry up,' or 'Go to the rear, boys.' About thirty men were packed like sardines in a gully which afforded a slight protection, better than nothing, and I was saying to them, 'Get up, boys, get up and go to the rear,' and pointing with my sword in the proper direction, when I was startled by the report of a musket close to my ear. I turned quickly and saw Corporal William H. Brown of Company B of the '11th' holding his smoothbore, from the muzzle of which fresh smoke was issuing. I didn't understand the conditions and in sharp tone spoke 'What on earth are you doing, Brown?' He replied, as he pointed at an officer in a gray uniform on the ground, 'That Captain was aiming his revolver at you when I fired.' I never heard of a similar occurrence in the war and think my antagonist was in delirium from pain and knew not what he did. At the memorable reunion in Gettysburg in 1913, Brown mentioned this affair to a group of the 'old boys' of both armies and an ex-Confederate said, 'That was my Captain.'" [Memoirs]

One of the staff arrived, and stated that a brigadier-general had decided to establish a new line of battle about a mile in the rear, but was unable to find his regiments, and delivered an order for the ranks to return at once to that point. The men were very indignant, because they wished to enjoy that rest which is so precious to every soldier, — a sleep upon the field which they had won by their bravery; and an officer[16] said, "Tell the general, that, if he will come to the front, he will find his commands with their colors; and, if he was not such a d—d coward, he would be here with them." They groped their way through the obscurity of the darkness, and passed by the first line, which was posted half of a mile from the Emmettsburg Road; and many of the troops were resting their rifles upon the rail-fence, and awaited an attack from the foe that was every minute anticipated; while the latter were dreading an onset by the pursuing forces. The halt was ordered for the night at ten, P.M.; a quietness that was rarely broken by the vigilant sharpshooters continued until morning; but the humble heroes of the day, not satisfied with their deeds of valor, requested leave from the general to go upon the field and succor their wounded comrades. The exigencies of the situation required their presence with the division: and the members of a small detail from each regiment took the canteens of those who slept, and carried the precious water, for which there was a universal cry, and bore the suffering to the hospitals in blankets and upon muskets and rails; while the chief portion of the ambulance corps was secluded in safe positions. Squads of rebels, who wandered over the plain upon a similar mission, strayed inside of the pickets, who captured them; but released one man, who said, "I am your prisoner, if you say so; but I am

____

16 Lt. Col. Porter Tripp, commanding the 11th Massachusetts. [Familiar Quotations] Blaisdell was absent on account of a chronic bronchial condition that flared up in June.

giving water to all that ask for it," and allowed him to continue his philanthropic labors.[17] A stretcher-bearer was badly wounded; and some surgeons expressed great surprise, and seemed to speak in terms of censure, because, unlike the majority of his rank, he had performed his duty upon the field, and incurred the dangers which were incident to the same.

Among the few Southern politicians, who fought on the battle-field for the diabolical treason which they had inculcated in the National Congress, was Barksdale of Mississippi, who led his brigade in the charge, and was mortally wounded within a short distance of the second line of batteries. He told the nurses of the regiment who were near him that he did not wish for any care, because he knew that he must die; and spoke of his family and home; and made only one allusion to the army, when he remarked, "Gen. Lee will clean you out of this place tomorrow." Major-Gen. Sickles, who was esteemed for his fearlessness by the corps which he commanded, received a severe wound in the leg, which was amputated. The officers of high rank, who criticised in such strong and unqualified language his conduct in advancing to the front and fighting the enemy instead of evading the onset, and sought to injure his reputation with the army and the people, would have displayed more wisdom and patriotism if they had adopted his policy in this respect, and remembered the maxim, the "errors of forwardness are forgiven, not backwardness."[18]

---

17 At the 50th Gettysburg Reunion in 1913, a picnic was held near the Rogers house site. "In talking with Dr. J. R. Evans, Ninth Alabama Infantry, I was astonished when I discovered he was the Confederate soldier who was taken prisoner while giving water to foes and friends." [Familiar Quotations]

18 Sickles' advance to Emmitsburg Road has been universally condemned by historians. While Sickles is criticized for the crisis on July 2 and the ruin of III Corps, Reynolds is given a pass for his defeat on July 1 resulting in the

The number of killed and wounded in this contest was very large: more than one-half of the division was disabled; eight color-bearers of the regiment fell; while the flag passed from one to another, and was never lowered; and the company to which I was assigned [Company K], which had thirty muskets at the commencement of the action, lost nineteen men by the bullet, seven of whom died of their injuries.[19] A part of the line upon the right had been fortified, and breastworks were constructed at other points during the night.

The rebel artillery opened with the dawn of daybreak, at half-past three, A.M., upon July 3, and continued their fire with unusual accuracy for an hour, at the position which was held by the left centre. The third shot exploded a caisson in the battery which was planted upon the left of the regiment; and fragments of wheels, and the woodwork, balls, and shells, ascended in a cloud of smoke and flame about one hundred feet into the air, and reminded me of the pictures which represent the eruption of a volcano. The division marched to the rear at eight, A.M., and was ordered to "ground arms" in the forest; and remained in a state of readiness to move at any point which might be assailed. Rations were issued, and greedily devoured; and no one who perceived the stillness that ruled at ten, A.M., would have imagined that two large armies confronted each other with the deadliest weapons of modern warfare in their hands. The concentrated batteries of the enemy opened at one, P.M., and shook the earth for an hour and a half with the terrific cannonade, — *Whose roar embowelled with outrageous noise the air.*[20] Lee

---

crippling of I and XI Corps.

19 Humphreys' 2/III division, with a battle strength of 4,924 had 2,092 casualties (43%). Carr's Brigade lost 790 out of 1,718 (46%). The 11th Massachusetts lost 129 out of 286 (45%). [GSL]

20 *Paradise Lost*, Boot IV (Milton) – referring to Hell's cannons.

had once more massed his infantry, and determined to make another desperate effort to pierce the left centre; and the division double-quicked to support those who occupied the earthworks in the front. The adjutant of one regiment, who noticed that his weak horse could not move as rapidly as the troops, dismounted, and ran to the scene of action, while he waved his sword with one hand, and led his steed with the other. The lines were formed at half-past three, P.M., and rested upon the ground, about six paces apart, during the conflict in which the fearful assault was triumphantly repelled; and Pickett's division was actually "cut to pieces" with spherical case-shot, canister, and lead.

"Grim-visaged war"[21] had suddenly appeared upon the field in which the division was aligned: the peach-orchards, flower-gardens, plats of green grass, and the golden harvest, pleased the eye, while the ear was entertained by the cackling of hens and chickens, and the squeals of pigs in the neighboring sty. The range of the rebel cannon was deadly exact; and different shells struck six men who occupied in succession the same place in the ranks; and the houses, barns, cellars, and yards were crowded with the wounded soldiers, who received accessions to their numbers during every minute. "Look out for that solid shot; don't stop it!" exclaimed a lieutenant, when the ball was rolling upon the ground towards the brigade; and a group of men in each regiment rose up, and left an interval through which it passed; but the spherical case-shot, which scattered scores of cast-iron bullets when it exploded, could not be avoided in this easy manner, and was very destructive. The Whitworth guns[22] threw their bolts a long distance; and the reports, unlike those of other cannon, could not be heard; and the peculiar humming of the shot

---

21 *Richard III*, Act I, Scene i (Shakespeare).

22 English-made, rifled, "snipers" cannon that could attain high accuracy at long ranges of 800 to 1,000 yards.

would be the first intimation of the discharge. Two soldiers in front of me were wounded by a piece of a gun-barrel, and others were lacerated by spikes. An artillerist, who was besmeared with blood, limped to the rear, and caused much laughter by his original and frightful oaths. The most amusing spectacle that I witnessed was a frightened brigadier-general, who sat in a wheelbarrow near a fence, dodged the missiles which did not come near him, and seemed to shrink to about one-third of his natural size.

"Lie down!" "Lie down!" was the invariable order for those who were not engaged with the enemy; and at one time, when two rebel caissons burst, Kearney's old division, which had been invisible, jumped upon their feet in front, uttered loud cheers, and then disappeared, apparently into the bowels of the earth.[23] It rose again, when three thousand prisoners of the assailing horde were captured;[24] and hundreds of hands pointed towards them before they were discerned by the brigade; and the soldiers [of Carr's brigade] turned their backs upon the foe in the midst of the shelling, as they gazed at the force which followed the roads to the rear with their colors. The black clouds overhung the sky during this fierce encounter; but the sun burst forth when the brilliant victory had been won, and cheered the wounded with its enlivening beams.

The rebels were dispirited by the repulse upon the 2d and 3d; called the plain a "slaughter-pen;" and declared that further fighting was useless; and some, who considered Jackson their

---

23 Birney's division was in front of Carr's brigade. The reserve line "formed in mass by battalion" and was in "support of the left of the Second and right of the First Corps." [OR XXVII/1, p.536]

24 Official figures place the total number of Confederates captured in Pickett's charge at around 3,700.

"very heart of hope,"[25] mournfully said, "We have not got Stonewall with us now." They related the following incident regarding Armistead, who commanded a brigade, and was killed in the unsuccessful charge. He skulked behind the trunk of a poplar-tree, in one of the battles before Richmond; and, as they advanced upon the open plain, several men who disliked him shouted, "There are no poplar-trees to get behind now:" and he replied to their taunts by saying, "Before this charge is ended, you will wish that there were some poplar-trees here." Some fields upon which the wheat flourished became the centre of conflict; the spires were trampled into the earth, and it was impossible to find one that was standing. Details were employed upon the forenoon of the 4th in burying the dead, and relieving the wants of the wounded, many of whom had remained upon the field nearly forty-eight hours, and were exposed to the perils of the sanguinary encounter which took place over their bodies. The rebel sharpshooters fired at all the fatigue-parties, and often shot at those who sought to alleviate their own comrades, that languished upon the ground within the limits of the Union lines, and could not be assisted by their friends. The supply of food, from some unknown cause, was deficient in the field-hospitals; and an application was made to the enlisted men in behalf of the wounded, and every soldier contributed liberally from his scanty store of rations. Experience in battle soon proved that the weapons manufactured in the United States were superior to those which were imported from foreign countries; and one regiment in the brigade, that bivouacked near a stack of several thousand arms which had been collected upon the field, threw

---

25 *Coriolanus*, Act I, Scene vii (Shakespeare).

aside their Belgian rifles, and selected those of the Springfield pattern.[26]

A member of a Pennsylvania regiment, who was at one stage of the conflict skirmishing upon his father's farm, near the house in which he was born, while the enemy held a position at the barn, refused to be relieved from his post of duty when the company was ordered to rejoin its command. Little did he dream that the strong arm and loyal heart which had contended against the foes of his country in the solitudes of Virginia would one day be required to attack them amidst the familiar scenes of his youth and home, and battle there with a courage which could never falter. All knew at noon that Lee had retreated; because the bands, clerks, and other non-combatants, arrived from the rear; and strains of music, intermingled with cheers, resounded along the lines from Wolf Hill to Roundtop.

The citizens, who deserted their houses when Lee approached, returned with their large families of small children, in haycarts and similar vehicles, which were followed by the horses, cattle, and swine which they had wisely taken away with them, and found in several instances merely a pile of bricks, and some charred wood in the cellar. Although a few of the inhabitants manifested a strong sympathy, and said, "Destroy our property, but drive away the rebels, and we are satisfied." Gen. Hays, a gallant Pennsylvanian,[27] who fell in the Wilderness, asserted in my hearing, that "the people who live on the border, in the vicinity of Gettysburg, are as base traitors as can be found

---

26 1/2/III brigade had a wide variety of weapons at Gettysburg:. .58 Springfield rifle (1st Ma.) .69 Springfield smoothbore (11th Ma.) .557 and .58 Enfield or Austrian rifles (11th NJ, 16th Ma.) .54 Austrian rifle (26th Pa). [GSL, 204]

27 West Pointer Alexander Hays (3/II) was one of the most outspoken of Union generals. He was aggressive to the point of recklessness, which some attributed to swigs from his flask. Original text says "Hayes."

in Virginia." Another officer from the same state remarked to me, "These Dutch farmers care for nothing except their cabbages; and, if they can make money out of Lee's army, they don't care how long they stay here." These tight-fisted miscreants, taking advantage of the necessities of the wounded, obtained a dollar for a loaf of bread or quart of milk; named a price for water and bandages; and, in the absence of most of the ambulances, conveyed them in their miserable wagons from the hospitals to the railroad depot, and demanded the most exorbitant amounts for their services. The clergymen and other prominent civilians of Gettysburg published a card in the newspapers, and boldly denied the truth of statements of this character; but I throw into the scales of justice the unbroken testimony of sixty-thousand soldiers of the Army of the Potomac.

The forces of Lee, which had been so recently flushed with the thought of Southern independence and the hopes of plundering the great cities of the North, retreated upon the night of the 4th from their earthworks, and abandoned thousands of the wounded, who were placed in the houses of the people upon the roads. The cavalry, led by its gallant commanders, at once commenced the pursuit; harassed the flanks and rear-guard of the enemy, and captured the trains and wagons: but the movement of the infantry was delayed, and the corps listened to the dim reports of the flying artillery of Buford, Kilpatrick, and Gregg,[28] but did not march for three days, although it was under orders "to be ready to start at a second's notice." The bugle at midnight awoke the soldiers, who were sleeping upon the huge rocks in the woods; and the troops, binding cords and straps around the legs of the pants to prevent chafing, advanced, and marched upon the road to Emmettsburg at 2½, A.M., of the 7th. A general and

---

28 The commanders of the three Union cavalry divisions.

certain mounted officers, who always procured government animals when a conflict was imminent, rode again upon their private steeds, which had been brought from the rear; but the majority considered that the most valuable horses were required at such a time, when the gain of a few seconds might change the history of an engagement. A brigadier remarked, in alluding to this fact, "I had two horses shot under me, and lost $1,200 at the battle of Gettysburg; but, if I had possessed twenty, that number would have been needed to keep in their places the cowardly ______ regiment," which was composed of ill-disciplined foreigners. The trees and houses for some distance bore the scars of the battle: many breastworks of rails and earth had been constructed to shelter the rebel lines; pits had been dug; and structures of small stones had been erected for the sharpshooters; and the right flank of Lee's army was protected by a formidable field-work, which had been ingeniously covered with branches, bushes, and transplanted savins, to conceal the troops and batteries, and deceive the distant Federal observer.

When all the facts attending this battle are fully understood, the historian will award the highest praise to the courage of the rank and file and the skill of the subordinate officers, and ascribe to Gen. Meade a very small degree of the honor for this decisive triumph. The conflict of July 1 was fought during his absence: the first corps captured a large number of prisoners; but the death of the accomplished Gen. Reynolds, and the re-enforcements which arrived for the foe, enabled Ewell to force the Union troops from their position, and drive them through the streets of Gettysburg. Gen. Howard had posted a division of the eleventh corps in reserve at Cemetery Hill; and this officer, assisting Gen. Hancock, who had the sole command, together with Generals Warren, Buford, and others, who noticed the great natural strength of the ridge, formed the divisions of the various corps upon the right and left of it; and thus the wise selection of the battle-field, a matter of the highest importance, which requires the exercise of the finest military judgment, was the result of a defeat. A part of this line, which these brilliant officers established while the commander of the army was several miles in the rear, was changed upon the 2d by Gen. Sickles, without any order; and the enemy for two days vainly assailed the gallant forces that held the original ground. During the gigantic struggle, Gen. Meade neither attacked the rebels, nor pursued them when they were completely shattered and had fled in confusion, but acted solely upon the defensive; and his able subordinates and their brave soldiers sowed, while he reaped, the harvest of martial glory which was produced by their successful labors upon the plains of Gettysburg.

Chapter XIII

Williamsport to Culpeper Court House

THE CORPS marched daily, from [July] seventh until the twelfth, through beautiful scenes which have been described, and hourly received the sincere welcomes of the loyal citizens of Western Maryland.[1] Severe storms frequently occurred, and affected in certain places the condition of the roads to such an extent that the wheels of the artillery and long trains of wagons made them impassable for the infantry that usually moved across the fields. A general [Carr], who sought to shield his acts of cowardice in the presence of the enemy by a display of arbitrary authority upon the march, stationed his staff to guard the bridges, and compelled the men to wade through the streams which often intersected the pathway, without allowing them any time to remove their shoes or clothing. This, with the exception of the infernal conduct of a commander who exposed the soldiers to the deadly rays of the Southern sun, or marched them for hours without a halt, is the surest mode of torturing them, or exhausting their energies; because the feet are quickly blistered, and a

---

1   VI corps pursued the Confederate retreat toward Hagerstown, but on July 6 it was determined that the direct pursuit via Fairfield Pass was contested: "Accordingly on the morning of the 7th, the whole Army moved by the flank routes through the Catoctin Mountains by Hamburg and the High Knob passes, and by way of Frederick." [Humphreys GR, 3] Thus, the Union pursuit march to Williamsport was appreciably longer than the retreat of the Confederates.

    Maj. Gen. William French was the new III corps commander. Humphreys became Meade's chief of staff, replaced by West Pointer Henry Prince at the head of 2/III division. Prince had served with both Grant and Blaisdell in the 4th U.S. Sickles recommended Blaisdell for division command based on his performance at Chancellorsville, stating that he had "served with distinction and fidelity in all the campaigns of this army, and is an officer of great merit." [OR LI/1, 1036-1037]

lameness ensues which cannot be healed for a long period. The Seventh New-York Regiment, which was encamped at Frederick City, and was the first force of militia that had been seen by the troops that performed the fighting, was greeted with derisive shouts by the veterans that belonged to the same city and State.[2]

Suspended to the limb of a tree which grew near the town was the body of a spy that had been hung by Gen. Buford, who acted promptly in this matter, without waiting for orders from the authorities at Washington. The collectors of relics had stripped the bark from the trunk, and taken from his person every rag except the portion of his clothing which was firmly held around his neck by the tent cord; and most of the old soldiers recognized him as one who had sold newspapers and maps to the army.

A captain in the regiment took off his shoes, and gave them to a notorious skulker who had alleged this pretext, and evaded the battle of Gettysburg, so that the man would have no similar excuse for his cowardice in the next conflict. The division marched over the battle-ground of the South Mountain Pass[3] upon the ninth, when — *The shades of night were falling fast* — and the color-bearers, who carried the blue silk flags of the Excelsior Brigade that formed the advance, reminded one of — *A youth who bore, 'mid snow and ice, A banner with the strange device, — Excelsior!*[4]

"Here Gen. Reno fell," was the simple inscription upon a plain rock near the road, which marked the spot on which America lost one of her greatest officers; and it is a melancholy

---

2 The 7th New York mentioned here is the militia regiment from Manhattan that had been mustered for thirty days during the invasion emergency.

3 The battle of South Mountain took place September 15, 1862, two days before Antietam.

4 *Excelsior* (1841) by Longfellow.

fact, that he was killed by the excited members of a new regiment who delivered a volley in the night.[5]

Upon July 12, the Army of the Potomac confronted Lee, who had concentrated his troops at Williamsport: the rolls of the companies were called once in two hours, and the usual arrangements were made for the battle which was every moment anticipated. The correspondents of the press misrepresent the facts nine times in ten when they assert that veterans are anxious to fight; but upon this day the soldiers who bore muskets wished to hear the commands, "Take arms," and "Charge," because they knew then, what is conceded now, that it would have captured all the cannon, *matériel*, and men from the enemy, and finished the Rebellion without a hard contest or a large loss of valuable lives.[6] When I recall the emphatic language that was used by rebel prisoners who were subsequently taken, refugees, civilians who were seized and detained to prevent them from communicating this information, and Union soldiers who escaped from their lines or were released and exchanged, I boldly state that nine-tenths of the officers and men of both armies would assent to this startling proposition, because I never heard one of them deny it. Lee had exhausted immense quantities of ammunition in the terrible combat at Gettysburg; many of his caissons and magazines did not contain a cartridge; and his horde could not withstand any onset. His men, disheartened by the knowledge of this fact and their heavy losses, wearied by the anxiety and severity of the hurried march upon the retreat, and unable to ford the

---

5   Gen. Jesse Reno was killed at the battle of South Mountain a year before while leading IX corps; accidentally shot by a startled soldier in the 35th Massachusetts. [Priest, 216, 218]

6   Humphreys wrote that "A careful survey of the intrenched position of the enemy was made, and showed that an assault upon it would have resulted disastrously to us." [Humphreys GR, 7]

Rappahannock,[7] which separated them from the base of their supplies, earnestly prayed that they could touch the soil of Virginia before the victorious Yankees arrived. The national soldiers, thoroughly equipped and furnished with sufficient ammunition; animated by the glorious triumphs of Gettysburg, the surrender of Vicksburg, the repulse at Helena, and the success which crowned the cause in every section of the country;[8] knowing the perilous circumstances of the disorganized mass in their front, and that a battle fought at this point would prevent an almost endless tramp, besides numberless conflicts in the disagreeable wilderness of Virginia, — wished with a united voice to be led to the work of carnage.

The mountains, which the archangel Michael, — *Of celestial armies prince*,[9] — wielded with such supernatural power when he crushed the hosts of Satan and Belial, were potent weapons in the all-powerful hands of Nature to assist the Union columns. The rain daily surcharged the springs that bubbled in the forests of oak upon the heights, and sent upwards slender clouds of vapor to stand in the air like sentinels, and point out to the national soldiers that were marching in the valleys, the abodes of their allies; while a thousand overflowing rivulets rushed down the steep sides of the lofty hills, enlarged their banks, removed the bridges and works of man which attempted to check the currents, and poured their waters into the Potomac until they had placed a stronger barrier than redoubts of earth or forts of stone in the rear of the "armed files" of treason, who were held day

---

7   The Confederates were backed up against the Potomac River at
    Williamsport (not the Rappahannock).

8   The besieged Confederates at Vicksburg, Mississippi, surrendered to Grant
    on July 4. The same day, the Union garrison at Helena, Arkansas, held off a
    Confederate attack.

9   *Paradise Lost* (Book VI).

after day upon the fields of Williamsport, and threatened to ingulf them whenever they fled before the avenging bayonets and rifled ordnance of the Northern forces. Gen. Meade, disregarding the wise advice of the heroes of Gettysburg, the fearless officers of the cavalry, and the generals that have been mentioned in terms of praise in the preceding chapter, read the bombastic address which Lee posted upon the walls of Hagarstown,[10] listened to the counsels of the timid and irresolute, was "afraid to strike." The stream subsided, and the golden opportunity was lost forever. The brigade advanced upon the morning of the 14th, and occupied the breastwork which had been erected by those who were posted in the extreme front; while three lines of battle marched over the bluff without opposition, and resembled in the distance the waves that roll over the ledges of a "rock-bound coast." A few slight pits for the infantry and sharpshooters were encountered, but they had been hastily constructed for show, and not actual service; while a number of men of straw appeared to be guarding the deserted ground. Deep gloom pervaded the army as soon as it was ascertained that Lee had been allowed to escape destruction; and, so eager were the soldiers to attack the enemy at this point, the reports of Gen. Kilpatrick's[11] cannon at twelve, M., produced cheers of exultation. For a long time the most awful curses were uttered in connection with the names of Meade and certain generals who opposed the assault. Six months after this shameful failure, I heard the shouts of some men, "Who voted against the attack at Williamsport?" "The drunkard _____!" "The traitor _____!" and noticed one of these obnoxious corps commanders, who was reeling to and fro upon his horse.[12]

---

10 Lee posted his "Proclamation to the People of Maryland" during his 1862 invasion, exhorting Maryland's citizens to support the Confederate cause.

11 Commander of the 3rd Cavalry division.

12 Blake is referring to the new III corps commander, William B. French. A

Several hundred lank and careworn prisoners, more than one-half of whom had no shoes, passed by the bivouac under guard; and one of them remarked as he pointed to a negro who was arrayed in the rebel uniform, "That is a Georgia cotton-picking nigger who would bring sixteen hundred dollars; but I will sell him to you for a loaf of bread."

The troops in the field diminished rapidly from losses by battle, exposure, and desertion; and a division, which was composed of the garrisons of forts at Baltimore, Washington, and Harper's Ferry, that had never seen a skirmish, contained six regiments with seven thousand men,[13] while the forty-two regiments of the veteran divisions of Hooker and Kearney presented for duty about six thousand men. The brigade marched upon the 15th across the field of Antietam: the soil which had been fertilized with the blood and bones of the slain bore bountiful harvests of wheat and corn; and the peaceful yeoman gathered the life-preserving grain upon the spot where, ten months before, death wielded his terrible sickle. A portion of the forest, which had been felled upon the crest of the mountains that towered above the battle-ground, formed an open space which the people called "McClellan's look-out," because that general viewed from this commanding height the conflict which raged beneath him. Although the foe was south of the Potomac, and there was no necessity for a forced movement, the corps was

---

bare majority of corps commanders opposed the Williamsport assault in a council of war on July 13, including French, on virtually his first day on the job. Meade delayed the attack one day and the Confederates escaped across the Potomac. President Lincoln was as upset as Blake at the failure to attack. [OR XXVII/1, 92] For a discussion of Meade and the council of war see Trobriand, 521-525.

13 The well-stocked division was the new 3/III division, command of which would eventually be given to Joseph Carr. Many veteran members of III corps marked French's assumption of command as beginning the decline of the III corps "as we understand it."

marched for seventeen miles, at the utmost rate of speed, from seven, A.M., till two, P.M.; when it halted for the day in the open field which was enclosed by pleasant woods that were reserved for the use of the generals and their staffs and horses. The sun diffused rays of fire; many raved in the delirium of its deadly stroke by the roadside; and some surgeons rode by the unfortunate victims without proffering their services, because they belonged to another command.

The brigade reached the ruined structures of Harper's Ferry, which nestled in the midst of the most picturesque and romantic scenery, and crossed the Potomac at eight, P.M., on the 17th, upon a pontoon bridge that was supported by twenty-five boats, and the wire bridge that spanned the Shenandoah. The first woman that I saw upon the southern bank repeated several times the characteristic wish, "I hope you will all get bullets in your heads;" which elicited from the soldiers a general reply, "How natural that sounds in Virginia!" No property was protected [by the rebels] in the States of Maryland and Pennsylvania, and the loyal citizens never uttered a murmur about the conduct of the men who conquered the foe at Gettysburg; but guards were placed upon the houses of rebels and guerrillas as soon as the Union forces crossed the river. A vague order was issued, allowing the soldiers to "take the top rail;" and a liberal construction was put upon this command, the qualifying word being considered a relative term; and each one was seized until the whole fence had disappeared, and the bottom became the "top rail" of the glowing campfires. The corps held the right, and daily advanced along the base of the Blue Ridge,[14] which separated the two armies, until Manassas Gap was occupied on

---

14 Carr's report indicated that the brigade was at Hillsborough July 18; Woodgrove July 19; Upperville July 20; Piedmont July 21; and moved up to Mannasas Gap by July 23. [OR XXVII/1, 546]

the 23d. All were inspired by the beauty of the mountains, the heights that receded at the gap, the variegated forest which adorned them, and the exhilarating atmosphere; but an ugly and degraded race of Virginians lived upon the slopes, waylaid the stragglers, and murdered the weary soldier who slept in the cabin in which he had been insnared by their hypocritical welcomes. The troops were always stupidly placed in the cleared sections of the country; and a citizen of one of the villages said that the rebel scouts upon the hills easily counted the divisions and noticed their positions, and repeated some of their statements, which showed an accurate knowledge in these respects. Many cripples were seen in the towns, who had lost their limbs while they were fighting against the national flag; and they invariably stated that they had been injured by reaping or farming machines. The blackberries were abundant; and the field which bordered upon the roads were covered with soldiers searching for them, whenever a halt was ordered. The people with scarcely an exception were rebels; and, while money would not tempt them to sell food, a small quantity of coffee overcame every scruple; and certain commissaries and similar officers made large profits by illegally selling the government rations. A general[15] gratified his tyrannical disposition by sending the pioneers in advance of his command to cut down and destroy all "foot bridges," so that the men would be compelled to wade through the numerous streams that intersected the road, and endure the suffering that always followed; while the scene highly entertained him and his staff. A woman in a village complained that a certain general treated her worse than the privates of his brigade; but cheating ignorant people in making change, or obtaining baskets and dishes by promising to return them when their contents had been

---

15 Probably still referring to Joseph Carr – he was not promoted to division command until October 5th.

consumed, were laughed over as splendid jokes at his headquarters.

The column moved five hours upon the 23d near the Manassas-Gap Railroad, which had been completely destroyed; and one resident near Piedmont made an ornamental iron fence of the rails which had been heated and bent in the centre. The utter depravity of the Southern slaveholder was daily revealed: a man who stood at the gate of his house, in reply to the question, answered, "There have been no rebels in this place within six months;" while his wife, who was in the kitchen, said, "A portion of Lee's army passed the day before yesterday." While the column was fording a broad stream that was knee-deep, a general (for whom, viewed as an officer or man, no one entertained any respect) vainly ordered the soldiers of his command to march in another place, and shouted to an officer as he pointed to a hole in the road where the water was four feet in depth, "Lieutenant, you go through there." No delay would have been allowed; and this lieutenant, knowing that the rations and ammunition of his company would be ruined if this useless order was obeyed, did not deviate from his course, and, by refusing to walk through cold, soon found himself in hot water, and was placed in arrest. A court-martial convened three weeks after this event; and although the general committed perjury, and testified among other falsehoods that, "the brook in the deepest part of it was not six inches deep," witnesses of inferior rank, but superior courage, honor, and veracity, contradicted his evidence; and his chagrin can be imagined when the subaltern returned to duty, and received no punishment.[16]

---

16 The lieutenant "in hot water" was Henry Blake, once again at odds with Carr. The court-martial was held in Warrenton on July 30 and found Blake guilty of disobeying orders but not guilty of "an insolent tone or manner." He was sentenced to "be privately reprimanded by the General commanding the Brigade." [Mutiny, 22-23]

The corps relieved the cavalry at Manassas Gap, and the rebels held possession of a part of it; while the rear-guard of Lee's army, which had marched from Winchester, passed from the [Shenandoah] valley to Culpeper Court House. Their skirmishers deployed at four, P.M., upon the crest of a high hill in front, which was the key of the position, and from which they were quickly driven, when the Union lines, more than a mile and a half in length, advanced.[17] A number [of rebels] purposely remained to be taken prisoners; and one of them said, "I am all right now." The soldiers in the extreme front moved forward with their usual coolness, picking and eating berries, and loading and firing their muskets as they clambered up the heights. The eyes were dazzled by the loveliness of the view from this point: hill rose above hill; the mountains changed their hues from green to the lightest shade of blue, until they became invisible; and the fields of wheat, with their rows of oats, looked like a vast network of gold in the valley through which the Shenandoah flowed. The trains of the enemy could be discerned in the distance upon one side of the mountains, while the Army of the Potomac was marching towards the gap; and Gen. Meade said in a tone of confidence, "We have got them now; to-morrow we will attack them." A corps commander, who was drunk,[18] and scarcely able to retain the seat upon his horse, rode along the lines, accompanied by most of his staff, including the non-combatants; and remarks like these arose: "There are no rebels

---

17 The battle of Wapping Heights, July 23, resulted from III corps being ordered to cut off the retreating Confederates by forcing the pass at Manassas Gap, defended by Walker's Brigade of Anderson's Division. Carr's brigade, including the 11th Massachusetts, was held in reserve, although the Excelsior Brigade suffered several casualties supporting 1/III division.

18 This would be William French. For another account of French's drinking and alleged incompetence at this time, see Trobriand, 530-531.

here;" "There won't be any fighting to-night;" and the men felt perfectly safe while they were present with them.

The soldiers rested upon a bed of rocks during the night; and the division with a squadron of cavalry and a battery made a reconnoissance in the morning of the 24th, discovered that the enemy had vanished from their position, and marched in pursuit to Front Royal, upon the road which was easily traced by noticing the newspapers, bags of ammunition, flour, and half-baked biscuits, which had been cast aside during the flight.[19] The wounded were un-cared for; the dead were unburied; and a faithful hound howled in the most mournful manner over the body of his master in the forest near Wapping Heights. A battery threw three shells at the head of the column at nine and a half, A.M.: the brigade at once filed to the right of the road, formed a line of battle in the woods, and waited for further orders. A small hill which rose abruptly in front interfered with the view in that direction: and, after the skirmishers had advanced, aides and other officers boldly rode upon the crest, and examined the ground; while a general [Carr] who showed base cowardice upon every occasion of danger timidly stood upon the slope, so that his eyes could barely see the position, and, repeating his ignoble conduct at Gettysburg, told the color-bearer of his brigade to "go to the rear." When the troops were ordered to move forward, this general was attacked by a disease which might be truly termed a case of indisposition; and the command devolved upon a lieutenant-colonel, who shouted the orders in a loud voice which might have been heard by the entire force of both armies. The first height was passed without opposition; and the men expected to receive a volley from the thick woods that crowned another hill which was beyond it, until the skirmishers reported that the

---

19 By the time 2/III division reached Front Royal, the Confederate army was safely beyond pursuit.

rebel cavalry were racing through the streets of Front Royal. When those in the rear learned this fact, the general, whose recovery had been as sudden as his illness, resumed his place amidst a thousand half-suppressed mutters and curses about the "coward" and "playing sick."

The column halted at this point an hour; retraced its steps at one and a half, P.M.; bivouacked near Piedmont; and encamped at Beverly Ford upon August 1, after a number of marches.[20] The rebel generals issued orders forbidding their soldiers to ask any questions concerning the towns through which they passed; and it is stated that Jackson always halted at the cross-roads at night, so that they would be unable to decide which route he would take in the morning. No such restrictions existed in the Union army; and the inhabitants were plied with inquiries, "What is the name of the next place?" or "How far is it from here?" One sagacious native of Salem, not wishing to be annoyed in this way, rendered a service by holding upon his knees a signboard, upon which all could read, "Warrenton, 13 miles." The regiment had marched four hundred and ten miles from June 11 to August 1, and seldom bivouacked two successive nights upon the same ground; and the rest at Beverly Ford was very desirable. The soldiers carried knapsacks when the grand movement commenced at Falmouth; but at this time most of them had a small roll, and did not possess more than one shirt, which was washed and worn again as soon as it was dry. The daily routine of camp-duty was resumed; and

---

20 Carr reported that the brigade advanced "through Front Royal halting on the west side for one hour and then went back towards Manassas Gap, reaching Markham at dusk of July 24, having marched about 20 miles. At 4 A.M. on Saturday, July 25, I marched 3 miles to Piedmont, halting near that place until 12 m., at which hour, … I proceeded with my command 3 miles beyond Salem going into bivouack on the Warrenton road at 5:30 P.M. On Sunday, July 26th left camp at 5:25 A.M., and marched to and through the town of Warrenton, and formed camp 3 miles beyond." [OR XXVII/1, 546-547]

the regiment furnished details to picket upon the Rappahannock near Freeman's Ford, where it was only forty feet in width and about two feet deep. The people who had transported their cattle, hay, and grain to the South to supply the rebel army, applied with their usual assurance to the commissary for rations to save themselves from starvation. They had sent away the few negroes who had not escaped to the North; and the able-bodied whites were fighting under Lee, so that laborers could not be procured to take charge of their estates; and the provost-guard was ordered to protect them. During the winter they had filled the ice-houses, which are usually built near the mansions of the wealthy, for the use of the sick and wounded in the hospitals at Richmond; and the luxury was confiscated for the benefit of the Union soldiers. A party of negroes ran away from Culpeper Court House; and, within half an hour after they had escaped across the Rappahannock, four blood-hounds, following their footsteps, appeared upon the opposite bank, and were shot by the pickets. Reconnoissances were frequently made, and several engagements took place at Brandy Station between the cavalry forces of both armies; and, upon August 4, the puffs of smoke from the cannon and exploding shells mingled with the clouds, and the reports of the artillery clashed with the reverberations of thunder during a severe storm. The Blue Ridge was unobscured by its drapery of vapor upon the following day. In the language of the residents, "The mountain took off its night-cap," and the rain ceased. An officer of the day directed a captain to examine that part of a river which was guarded by the division, and ascertain, if possible, the number of points at which it could be forded. The clothes of the officer were thoroughly drenched with water when he returned in the afternoon, and reported that he had waded in the centre of the stream nearly a mile, narrowly escaped drowning, and stated the results of his unforeseen method of sounding the Rappahannock.

The camps witnessed an affecting spectacle upon the 14th. The veterans of many honorable battles — the officers and men of Kearney's and Hooker's divisions of the third corps — contributed their pay for one day to purchase for Gen. Sickles, their gallant and disabled commander, a carriage, horses, and harness, as an expression of their respect; and, when the wounded returned from the hospitals, they would not be pacified until their names were added to the long list.[21] The good opinion of these brave soldiers — of one man who bore a musket, and had seen and admired his conduct at Chancellorsville and Gettysburg — was of far greater weight than the carping of generals who sat in their chairs of ease and safety at Washington. The cheers of such voices, and especially those from the ranks, will resound through future centuries, while the contemptible sneers at Gen. Sickles and the heroes of New Bern and Lookout Mountain are imperceptible.[22]

A squad of two hundred substitutes (there was not one conscript among them) was assigned to the regiment after tattoo on the 23d [of August]; and the utmost vigilance was required to retain them within the limits of the camp. A more motley crowd was never inspected. Every nation and occupation was

---

21 Sickles was on the banquet circuit in upper New York state at this time. [*New York Times*, August 13, 1863]. The gift was not presented until mid-October, while III corps was on the Bristoe Station campaign. "It must be acknowledged that this reception was not only a manifestation in honor of the old corps commander, but also a protest against the successor given to us." [Trobriand]

22 New Bern (March 14, 1862) was a Union victory in North Carolina where Burnside led the capture of the key port city; Lookout Mountain (November 24, 1863) was a Union victory by troops under Hooker. All three – Sickles, Burnside, Hooker – have long since been the subject of "contemptible sneers" which III corps veterans would feel unwarranted.

represented: thieves, organ-grinders, garoters, and New-York [draft] rioters, formed a majority; and all, with a few exceptions, intended to desert at the first opportunity, to obtain another bounty.[23] Twenty or thirty had been daily tied up by the thumbs during the voyage; some had been shot while they were swimming to the shore; and others, by a system of general pillage, accumulated amounts that exceeded two thousand dollars. Many had deserted from various branches of the service, and understood the manual of arms and the company movements. It had been announced that five bounty-jumpers would be shot in the fifth corps upon the 29th; and the day was awaited with the deepest impatience by the officers, who could not be held responsible for a lax state of discipline if the villains were pardoned by the President, and by the substitutes who made preparations to leave if the execution was postponed. The miserable wretches were marched to the ground where five graves had been dug two hours before the fatal moment: each man gazed upon his last resting-place, and then returned to the prison. The fifth corps was formed under arms upon the field, besides squads of conscripts who were under guard; and most of the third corps were present as spectators. The band of the regiment played the "dead march"[24] while the procession was moving to the scene; and each prisoner, with his hands manacled behind him, walked in the rear of his coffin, which was carried by four soldiers, and placed in front of the grave. Two were Jews, and two were Roman Catholics; and the rabbi and priest who accompanied them had a dispute about precedence, and urged their respective claims upon theological tenets; but the

---

23 The 11th Massachusetts roster records 187 draftees (or their substitutes) in July and August 1863. Of these, 58 (31%) eventually succeeded in deserting. [MSSM/1]

24 The "dead march" is from Handel's oratorio *Saul* (1738).

commander of the provost-guard viewed the subject in a military light, and decided the novel question by allowing the rabbi to walk first, because his faith was the oldest and outranked the other. The last solemn rites were celebrated; each culprit sat upon his coffin; their eyes were bandaged; within a second the bullets from fifty muskets pierced them, and soon five mounds of earth covered their bodies.[25]

The orders to march[26] at sunset upon September 15 were so unexpected, that a wagon loaded with evergreen and boughs[27] for headquarters passed by the camp while the "general" was beating, and the soldiers were striking tents and packing up their effects. The column moved at half-past seven, P.M.; but a major-general [French] was intoxicated; great confusion prevailed in consequence of conflicting orders; and the division marched in a circle through the woods, hour after hour, until one, A.M., of the 16th; and actually halted for the remainder of the night, at the end of this most tiresome and needless gyration, within an eighth of a mile of the quarters which had been abandoned. It was a mile and a half to the ford at which the crossing should have been made; and a large number of officers and men could have pointed out the place without any difficulty, and avoided this over-exertion. The troops, at an early hour upon the 16th, forded the Rappahannock, which was knee-deep; and subsequently Hazel Run, which was hip-deep; and Gen. Prince, a most

---

25 This execution of five deserters from the 118th Pennsylvania was given front-page coverage by the the *New York Times*, September 2, 1863 and *Frank Leslie's Illustrated Newspaper*, September 26.

26 The II corps, with cavalry, had been "thrown across the Rappahannock on the 13th of September, driving the enemy's cavalry over the Rapidan. The army followed and took position about Culpeper Court House" [Humphreys GR, 11]

27 Pine branches were used in camp to cover the unpleasant smell of men living in close quarters with rare opportunities for bathing.

exemplary officer, who commanded the division, was placed in arrest because he allowed them to remove their shoes or boots when they travelled through the water. The enemy retreated south of the Rapidan; and the camp of the regiment was located, upon the 17th, about a mile in the rear of Culpeper Court House. The cold often interrupted sleep at this season, before the occupation of the winter barracks; and the fires would be surrounded by groups of shivering soldiers, two or three hours before twilight.

Chapter XIV

To Centreville and Back

CULPEPER COURT HOUSE consisted of deserted buildings with broken windows; empty stores; a few destitute natives; a jail, and similar institutions; and four churches, from which the pews had been removed to render them fit for occupation by the sick and wounded soldiers. The commanding heights, the Blue Ridge, and Cedar Mountain, which is known among the natives by the common family name of "Slaughter," which the disgraceful blunders of the battle[1] made very appropriate, rose in the front, and surrounded the city. The most stringent commands were issued to the members of the brigade to prevent them from taking the fences of a notorious rebel, an Ex-M.C., upon whose grounds the camp had been established:[2] a large force was detailed to guard them, and four written orders were read to the men upon this subject in the course of half an hour. The troops of Lee's army did not injure the property when they bivouacked in the same field; and I solved the perplexing problem regarding this singular conduct, when I ascertained that the wealthy owner was the father-in-law of one of the generals of the Union army.[3] The quartermasters and teamsters of one division pitched their tents and parked their wagons in a cemetery; and some of these

---

1 The Battle of Cedar Mountain was on August 9, 1862.

2 John Strother Pendleton (1802-1868) was elected as a Democrat to the U.S. Congress from 1845 to 1849. His estate was called "Redwood."

3 While there was indeed a connection between Pendleton and a serving Union officer, Blake misrepresents the actual situation. Pendleton had adopted the children of his brother-in-law (Philip Williams) upon the death of their mother in 1837, including Maj. Robert Williams: West Point 1851; aide to Gen. Banks 1861; 1st Mass. Cavalry 1862; Asst. Adj. Gen., Dept. of the South 1863. [Museum of Culpeper History]

unfeeling non-combatants levelled the mounds of earth to secure a better floor for their shelters. The health of a number of substitutes in this brief period seemed to be as frail as their reputation for honor; for some of them were crippled or unsound in an organ of sense; one was so blind that he was always piloted in the night by seizing the end of a musket, while a faithful comrade carried the other; black hair gradually lost its color, and the white head of a person too old for any service appeared; and two [substitutes] died of consumption within a month after their arrival.

The military position was unchanged until October 8, when the division was detailed for special duty, and marched to James City to support Gen. Kilpatrick during the skirmishes between the cavalry.[4] The rebel camp-fires burned at night with their usual brilliancy; the tents and shelters were not removed; the pickets maintained a strong force at the same fords upon the Rapidan; while Lee moved his army upon "circuitous and concealed roads,"[5] and intended to pass by the right flank, and rush to the strong position of Centreville. The observing eyes of the signal-corps, who were posted upon the summits of the mountains, promptly discovered this ingenious design. The infantry acted as a support for the cavalry; and general quietness reigned until three, P.M., of the 10th, when a brisk skirmish commenced that did not cease until night, and the enemy was completely foiled. In the mean while, the cars were loaded with stores, and sent to Alexandria; the wagon-trains were in motion; the main body of the army was already preparing to retreat across the

4 Kilpatrick's cavalry division was active during the month of October, skirmishing with Confederate cavalry. Official reports reflect a lengthy exchange between French and Prince as to whether the latter properly supported Kilpatrick on October 9.

5 The quoted phrase appears in the report of Gen. Lee dated October 23, 1863. [OR XXIX/1, 410]

Rappahannock;[6] and the division began to return before sunset, as the orders were explicit to avoid bringing on a general engagement. The troops filed into the fields near the road to bivouac for the night; and had barely fallen asleep before the march was resumed, and there was no halt until midnight. Most of the corps crossed the river upon the 11th, and every man uttered a yell while he forded it. The rebel cavalry closely followed the rear to pick up stragglers, while the [rebel] infantry was attempting to make a grand flank movement: and the contest became a race between two armies, which hastened, upon the routes that were nearly parallel, to gain the same point; and, although the Federal forces were encumbered by the trains, they won the position, and were only two hours ahead of the advance of the enemy. Some generals injured the service by placing their sons and relatives upon their staffs,[7] and sending orderlies to perform their duty, and carry important commands, when their lives were endangered by the battle; and the officers of the cavalry usually detailed for this purpose the most worthless soldiers that were mustered upon the rolls. One of this class, who could not speak English, delivered an order to me when I commanded the skirmishers that covered the brigade as it fell back from James City; and I was unable to interpret his jargon at a time when a deviation from the proper path involved capture and other serious results.

---

6 Humphreys wrote that by the night of October 11 "Lee's army was moving toward Woodville, some fifteen miles northwest of Culpeper … and this evinced a design on his part not to advance upon Culpeper Court house, but to turn our right," and since Lee "was already as near to Warrenton as we were, General Meade decided to move at once to that town."
[Humphreys GR, 15]

7 Meade and Humphreys employed their sons in this capacity; Carr, his brother.

The regiment was stationed at Beverly Ford upon [October] 12th, behind the earthworks which they assisted to build, while on the march to Gettysburg [four months previous], to prevent the enemy from crossing at this point.[8] The pickets of the rebel cavalry that were posted in groups upon the opposite bank hastily rode away as soon as Stuart was forced to retreat to Culpeper Court House; and their rapid flight was the first result of the national success. The division marched from sunrise to sunset upon the 13th, and was delayed during the night by the bad state of the road; and fires were built when these irregular halts occurred, and several miles of rail-fence were destroyed. The column rested at Greenwich only two hours; moved at daybreak with great caution: flankers marched through the woods and fields; and companies held the by-paths to protect those in the road against a sudden attack. The lines of battle were formed at two, P.M., upon the heights of Centreville, where the regiment was aligned for the eighth time during its various campaigns, which were termed by men "forward and back" movements.[9]

"We must pass through the crack of a door," Gen. Prince remarked in the morning; and the fatigued soldiers were urged to keep in reserve all their strength, because the safety of the army depended upon their promptness and power of endurance. These forced marches, which could not be avoided, caused great exhaustion; and many substitutes gladly straggled from the ranks,

---

8 Also on October 12, Meade, temporarily abandoning his belief that the Confederate infantry was flanking his right, sent V, VI, and II corps, with Buford's cavalry, back to Culpeper Court House to confront the Confederate infantry which (it turned out) was not there.

9 October 13th, III corps passed II corps on the march eastward toward Centreville. The latter, as the new rear guard, was attacked at Bristoe Station by Gen. A. P. Hill's corps. The Confederates were held off until nightfall and successfully withdrew to Centreville, with substantial and disproportionate casualties.

and concealed themselves until the rebel cavalry advanced, when they surrendered like willing prisoners. The natives of this section of Virginia did not appear to own any of the estates which they occupied; and most of the houses displayed a signboard, upon which was painted, "British property. Safeguard placed by Gen. Sigel or Meade."

The corps constituted the left of the army upon the 15th; and the division proceeded to Union Mills, where one company from each regiment of the brigade was detailed for picket-duty. The main body was posted at the ford and bridge that crossed Bull Run at this point; and the company of which I had charge halted at the base of a hill which commanded the stream, and upon which earthworks had been constructed by the troops of the right wing of Beauregard's force at the first battle in 1861. Three men and a corporal relieved a squad upon the crest; and a staff-officer who gave instructions, and the men that composed the old guard, said than no rebels ever molested them, and there was no necessity for unusual vigilance. When I reconnoitred the ground in the vicinity, and passed through the thin belt of woods which was two hundred yards in front of the outposts, I saw, at the distance of half a mile upon the broad plains of Manassas, a line of advancing skirmishers, which was supported by a battery and a regiment of Stuart's cavalry. The company and a few riflemen from the reserve were ordered to re-enforce the little command of the corporal. The first ball surprised their ears, and was followed by a halt; after which the principal portion dismounted; and every man in the rear held the reins of four horses that belonged to his comrades, who were repulsed in every effort which they made to surround the pickets; and a scattering fire continued for two hours. They then moved to the right, planted their battery near McLean's Ford, and surprised the pickets that were compelled to recross the river. The four soldiers who had been relieved belonged to a brigade that had recently arrived from Suffolk; and one of them entertained the listeners by describing the numberless battles in which he had participated upon the Blackwater [River].[10] When the bullets whistled over the crest, and I was watching the movements of the foe, they ran to the rear; but the story-teller was seized and thrust into the front rank, although he pleaded most earnestly for his release, and admitted that he had never seen a fight, and that his brigade had "done nothing" since it entered the service.[11] A sergeant, who ineffectually kicked a substitute that crouched upon the earth and refused to rise and discharge his musket, grasped him by the collar, and held him up: so that he was exposed to the fire of the enemy, until he brought his piece to the shoulder and pulled the trigger. The storm which arose in the night, and the tall wet grass, chilled the sentinels, who remained in the same spot, without any fires. In the morning, a chain of pickets, who sat upon their horses, extended from this point more than a mile to the rebel camp.

The subtle plan of Lee had wholly failed,[12] and no wagons or organized bodies of infantry were captured; but he was leisurely followed [by the Union army] when his forces were withdrawn: and the division was encamped near [Warrenton] Junction, about seven miles from Bealeton, upon October 30, while the right of the army extended to Warrenton. The corps commander [French]

---

10 The newly-arrived brigade had been part of the Union occupation force holding Norfolk, Virginia, south of the James River. Union and Confederate forces faced off at Suffolk the first half of 1863.

11 Possibly Corcoran's Brigade of XXII Corps, Dept. of Washington, whose duties included guarding the Orange and Alexandria RR. It was part of the 1/VII division when stationed at Suffolk. [Dyer/1, 332, 378]

12 The October 1863 Confederate movement around the Union right flank resembled the one of August 1862 that led to 2nd Bull Run. However, this time they were repulsed at Bristoe Station.

most unjustly favored the third division, that had never performed any fighting, by always placing it in safe positions;[13] while the first and second, which had fought the enemy again and again, were exposed to every danger. The roads were so narrow, that a single team obstructed the passage of those that were in the rear of it: and government property was summarily destroyed if it could not be removed; but the covetous sutlers were unwilling to adopt this policy, and often blocked the trains during the movement. Gen. Meade issued a just order, which restrained their privileges, and banished them from the army for a certain period. The rebels injured the Orange and Alexandria Railroad to the utmost extent; demolished the bridges, water-tanks, and culverts; and ruined the iron rails by placing them upon a pile of burning sleepers, the heat of which softened the centre, so that the ends rested upon the earth; and some were twisted around the trunks of oaks which grew near the embankment.[14] The track to Warrenton had not been disturbed; and the rain interfered, at certain points, with the work of destruction, which was imperfectly executed. The soldiers were now required to perform a new species of labor; and large details were daily furnished to grade the road and fell the trees, and cut sleepers or ties. The buildings and fences had disappeared; and the general barrenness which prevailed was occasionally relieved by the green spots of ground, in which the grain that had been scattered by the cavalry had taken root, and sprouted. Two dead horses, which were

---

13 Much of 3/III division had been led by French while at Harpers Ferry in mid-1863. It joined III corps after Gettysburg and was placed under Carr when French rose to corps command. They were called "lambs" by the veterans of the 1st and 2nd divisions of III corps.

14 A consequence of the Union army falling all the way back to Centreville was the destruction of about twenty miles of the railroad supplying their position around Culpeper. However, the U.S. Military RR had it repaired by early November. [Humphreys GR, 36]

respectively branded "C.S." and "U.S.," were stretched upon the field, near Bristow Station, with their heads a few feet apart; and all who witnessed them asked, "Where are the riders?" A rebel cavalry scout captured three unarmed soldiers who were wandering outside the picket, and ordered them to take down a fence which his horse could not leap; and each one seized a rail, dismounted him, and retraced their steps with the prisoner.

The army, like all travellers, "took an early breakfast" upon the morning of November 7, and advanced towards the Rappahannock in two columns; and the corps marched to Kelly's Ford, where it arrived at three, P.M.[15] The hill upon the northern bank commanded the position; and the third brigade of the first division[16] waded through the river while the batteries were briskly engaged, successfully charged upon the rifle-pits, and took five hundred prisoners.[17] A bridge composed of eight pontoons was immediately constructed over the stream which had been many times passed and re-passed, guarded and re-guarded, by the Federal army; and the brigade crossed at sunset, when the flashes of the rifles revealed the locality of every skirmisher, and the cannonading resounded from Rappahannock Station.[18] A Union captain was killed by a sharpshooter while he

---

15 The left column (I, II, and III corps) under French, headed for Kelly's Ford. The column on the right (V and VI) was commanded by Sedgwick. [Humphreys GR, 39-40]

16 Commanded by Col. Régis de Trobriand.

17 "This affair did not cost us a hundred men, and brought us in more than five hundred prisoners." [Trobriand, 549] Humphreys recorded the Confederate losses at Kelly's Ford as 5 killed, 59 wounded, and 295 missing. [Humphreys GR, 41-42]

18 The distant fighting at Rappahannock Station was more severe than at Kelly's Ford. The night attack by Russell's 1/VI division resulted in a significant Union victory: "103 commissioned officers, 1,200 enlisted men, [and] seven battle flags were captured." [Humphreys GR, 44] In the two battles "more than 2,000 irreplaceable seasoned veterans were gone — a

was in the act of giving some water to a wounded rebel who was moaning upon the field.

"You have got our winter quarters," exclaimed some of the prisoners, who stated that they were completely surprised, because they had often formed in line of battle to no purpose when the cavalry was reconnoitring in their front, and expected no unusual event when their pickets escaped to the support. The vent-hole of a cannon is always closed when the gunners are loading it, and a thumb-stall is generally worn to prevent the blisters which would arise if the piece became heated during an active engagement; and an accident seldom occurred. In this contest the person instinctively removed his thumb, which was unprotected; and a premature explosion resulted, by which three men were mangled. The enemy retreated across the Rapidan in the night;[19] and the division marched to Brandy Station without opposition on the morning or the 8th. Several negro servants that labored for rebel officers of high rank entered the lines, and brought with them the horses and overcoats of their masters; and one carried a large basket which contained the dinner and dishes of a brigadier.

Lee's army had constructed barracks for occupation during the winter in the dense forests in the vicinity of Brandy Station, and collected bushels of acorns for food. The [rebel] quartermasters issued clothing without any buttons, which were cut from the old and discarded garments, and sewed on the new by fortunate receivers of the butternut. They had gathered hundreds of empty tin cans, which had been sold by the sutlers

---

staggering loss, especially coming on the heels of the Confederate casualties at Gettysburg and Bristoe Station." [Mine Run, 28]

19 Meade relented on the pursuit that had caught Lee off guard at Rappahannock Station and Kelly's Ford, allowing Lee to withdraw south of the Rapidan. [Mine Run, 35]

when they were filled with pickles or preserved meats, and intended to use them as dippers; and many correspondents at once inferred from the labels that there had been illicit trade between North and South. The division occupied the camps which had been allotted to some brigades in Ewell's corps; and the log structures afforded comfort to the enemies of the builders.[20]

---

20 Meade's stock in Washington soared after his success at Kelly's Ford and Rappahannock Station. It quickly dropped when Lee once again slipped away untouched across the Rapidan. Secretary of War Stanton was so disturbed that he was even less gracious than usual: "Meade's timid performance became all too evident on November 11 when the secretary [Stanton] refused to receive an honor guard carrying the enemy colors captured at Rappahannock Station." [Mine Run, 40-41]

Chapter XV

The Advance to Mine Run

THE EXTENSIVE preparation for a decisive battle had been completed, and the pontoons which rumbled upon November 23 over the roads that led to the fords of the Rapidan were the forerunners of a general advance.[1] Marching orders were received at the unwelcome hour of midnight, but countermanded at daybreak when the regiment was forming its line; tents were re-pitched in the midst of a rain, and the mud held fast many wheels which must revolve to supply the necessities of the army. Upon the 26th, as the beams of the rising sun touched the wintry frost, and concealed the crests of the Blue Ridge in the clouds of its vapor, the corps abandoned their camps; and the proclamation that was printed in the newspapers was the only evidence which satisfied the soldiers that it was a day set apart for Thanksgiving.[2] Great enthusiasm was produced by the reading of the despatch of Gen. Grant, announcing the victory at Chattanooga.[3] The grand force moved forward, and was soon separated from any base or line of communications. The cavalry menaced the upper fords to deceive the enemy, while the infantry advanced in a different direction; and the corps (the third), followed by the sixth,

---

1 Meade decided to continue the campaign against Lee south of the Rapidan as a "prompt movement to cross the lower fords nearest to General Lee's right in three columns. … The plan promised brilliant success; to insure it required prompt, vigorous action, and intelligent compliance with the programme." [Humphreys GR, 50]

2 Sarah Josepha Hale had advocated for a national day of thanksgiving prior to the war with editorials in women's magazines and letters to Governors and Presidents. Union military successes of 1863 probably influenced Lincoln as well.

3 Battle of Chattanooga, November 23-25, 1863.

constituted the right column.[4] The brigade was posted in the advance, and passed by only four dwellings in marching fourteen miles, and halted in the woods about half of a mile from the stream, at a point which was styled, by the name of the nearest inhabitants, Jacob's Ford. The bluffs of the southern bank rise very abruptly one hundred or one hundred and fifty feet in height, and are well adapted for defensive operations.

The passage of a river by a corps, in the face of the enemy, is considered in military treatises one of the most difficult movements known in war; but the Army of the Potomac has performed this hazardous undertaking with success upon every occasion, while the rebels never attempted to effect it. The skirmishers of the brigade deployed from the forest, and marched to the ford, which is located at a sharp bend: a battery was planted to protect them; and the small body of rebel cavalry that witnessed these dispositions fled without firing a carbine. The pontoons were transported to the river, and placed in the water; and the skirmishers, supported by four companies of the regiment, immediately embarked in the boats, and clambered the heights, upon which the brigade-color waved without opposition.[5] Two incidents which fell under my observation at this time show the results of the negligence of some staff or general officers to thoroughly perform their duty before the movement commenced. The width of the narrow stream had been miscalculated, and there was a deficiency in the number of boats that were required, so that a delay of an hour occurred while the

---

4 The right column crossed the Rapidan at Jacob's Ford to establish the right flank of the Mine Run line at Robertson's Tavern. I, II, and V corps crossed downstream at Germanna Ford to extend the line south astride the Orange Turnpike and Plank Road. [Humphreys GR, 51]

5 According to Prince the skirmishers were from the 26th Pennsylvania and the crossing party in the boats was from the 11th New Jersey. [OR XXIX/1, 760]

pioneers were constructing a support of earth and logs for the southern extremity of the bridge. The men who had been drilled to execute this peculiar labor quickly anchored the pontoons at certain intervals in the swiftly flowing current, arranged and fastened together the planks of the trestle-work, with the regularity that characterizes the movements of a machine. The troops began to cross at three, P.M.; and discovered another oversight which was obvious at a glance, that the slopes were too precipitous for the passage of cannon and wagons; which were then sent to another ford that was two miles from this point, because there was not time to cut a new road.[6] A severe battle had been anticipated at the fords; and every eye looked upwards with earnestness during the most anxious moments of the day, when the skirmishers slowly approached the crest of the bluff. "Is the foe concealed behind the hills that frown upon us? Does he crouch in ambush in the thickets of the Wilderness beyond them?" These thoughts flashed through the mind; and the steps are slow; the musket is held with a firmer grasp; the finger constantly rests upon the trigger; and every object is scrutinized, because a single mistake might cause death or defeat. Thus the Army of the Potomac won its new position south of the Rapidan, before sunset, without losing a life; and Lee, who had massed his columns to resist the advance of the national forces at the points where the cavalry was making alluring feints, was chagrined to find that he had been outgeneralled, and his enemy threatened his rear and right flank.

---

6 Meade was already upset with the slow progress of III corps, which had been an hour late in starting. French blamed Prince, who he claimed was habitually slow to move, and failed to employ a guide after crossing the Rapidan. [OR XXIX/1, 739] French reported that he had sent a guide to Prince on the evening of the 26th, which contradicts his later statement that Prince had no guide after crossing the Rapidan.

Glorious visions of success already enlivened the hopes of the soldiers. Celerity of action, the concentration of the corps at Robertson's Tavern, which placed them between the commands of Ewell and Hill, would produce a conflict with the foe cut in twain, which must insure the victory. The brigade, enclosed by strong lines of skirmishers and flankers, still formed the head of the column, and marched in compliance with instructions upon a narrow pathway which led through the almost impenetrable Wilderness until the rebel pickets were encountered. They were speedily driven more than a mile; and the yelping of a wounded dog, an animal which always accompanies them, indicated the course which they had pursued in the darkness.

"Tell the division general that my skirmishers are scalping the devils like h—l!" was the verbal message which the colonel commanding the brigade[7] duly transmitted to announce his triumph.

"We are in the bowels of the enemy," remarked Gen. Prince, who ascertained that the troops had moved upon the wrong road, and were advancing to the fortified stronghold at Morton's Ford.[8] He ordered them to countermarch: a tangled and unbroken forest increased the gloom and weariness of the three miles which were retreated to a point near the ford, and the bivouac was established for the night. These precious hours that had been lost by the faithlessness of a corps commander delayed the whole army;

---

7 Blaisdell had just returned to brigade command after a seniority dispute with Col. Robert McAllister (11th New Jersey).

8 It does appear that Prince's 2/III division took a lengthy walk towards the enemy line on the late afternoon of November 26. The lead regiment's commander, Lt. Col. Robert Bodine, 26th Pennsylvania, indicated how far off course they were to the west: "[we] crossed Mine Run at dusk, driving away the rebel pickets, and went into bivouac at Tinsley's Mill." [OR XXIX/1, 769]

enabled Lee to unite his dissevered divisions; and the sacred cause of the country was frustrated for many months.[9]

The breakfast of coffee and hard bread was devoured before daybreak upon the 27th; and the lines were formed to resist any attack which might be made by the enemy that hovered in the forest. The march was resumed in the morning in compliance with orders from a corps commander; light skirmishes frequently took place with small squads of cavalry; and at noon there had been no junction with the main body, although Gen. Meade had labored unceasingly to secure this object. In consequence of the blundering oversight which has been noticed, the ammunition trains and the artillery were not present to assist the infantry; and the perilous situation in which the latter was placed can be discerned without a lengthy explanation. A long and vexatious delay occurred because another move had been made in the wrong direction; and the men, justly dissatisfied, rested in the road, and pulled down the fences to build fires, which were extinguished to prevent the [rebel] scouts from gaining a knowledge of the brigades by watching the thin clouds of smoke. A corps general, who entered the house of Jacobs upon the preceding afternoon, and remained in the rear while thousands of his victims were wandering in an unknown region, arrived; established his headquarters in another dwelling; and I did not see him again during the eventful day.[10]

---

9 The historian of 11th New Jersey supports Blake's version of a wrong road and a lengthy countermarch: "We had advanced about four miles when, from a mill in front, our column was again fired upon. Darkness had fallen, and it being ascertained that we were upon the wrong road — one that would lead us in front of the enemy's works instead of upon the flank — we fell back two or three miles and bivouacked for the night." [Marbaker]

10 Miscommunication between French and Prince continued on November 27th. French, at III corps headquarters, was expecting a report from Prince, who was stuck at the Widow Morris' farm crossroads awaiting return of his scouts. French took matters in hand, ordering Prince to advance on the

Although this officer was usually know as the personal friend of Gen. Halleck, justice demands that his shameful conduct should be fully described; because the failure of this finely conceived movement, in the opinion of the author and thousands of comrades in arms, was caused by one of the corps commanders, who discarded Mars, and served Silenus.[11] Habitual drunkenness had covered his face with frightful blotches, and destroyed his control over some of his muscles; the cheeks twitched convulsively, while the eyes and mouth opened and closed in a comical manner which would have insured the fortunes of a clown. The derisive laugh which an intoxicated fool always excites greeted his appearance; and I extracted from the numberless oaths and jeering remarks that were uttered at this time the following specimens: "Old blinky has got up at last!" "His horse is drunk again to-day!" "Here comes the old gin barrel!" "I should like to tap him!" "I hope the first cannon-ball that is fired will knock his head off." Discipline under such a sot is maintained by the ceaseless efforts of the subalterns and the undying patriotism of the men who mourn the absence of the gallant and noble-minded leaders of other days, who died amidst the storms of conflict. However, the coward, the traitor, and the drunkard of high rank may skulk from the scene in the decisive hour of the combat; but the veterans that never quailed are inspired by the dead heroes who mount again their war-horse, draw from their scabbards the two-edged swords, and advance in the charge.

---

right fork from Morris farm. This led Prince toward the Confederate line, resulting in the Battle of Payne's Farm.

11 Mars was the Greek god of war. Silenus was a lesser god associated with Bacchus, "a jovial fat old man, who usually rode an ass because he was too drunk to walk." [Hamilton, 40-41]

The brigade pushed forward in line of battle through the woods upon both sides of the road, and drove the rebel pickets, with their supports, two miles, until they reached a house which stood in an open field. The troops double-quicked at two and a half, P.M., from this point to Locust or Orange Grove, in which a sharp contest ensued between the skirmishers,[12] and steadily forced the enemy to rejoin the principal force which was posted in the road that ran to Raccoon Ford.[13] The regiment, and that upon its left, held a position upon the slope of a gentle ascent against an attack; and the rebels, who could not stand erect and face the shower of lead, crouched upon the earth, and sought the protection of the crest.[14] A part of another brigade gave way upon the right, and exposed that flank, so that the regiment was obliged to fall back to the cleared tract, where it re-formed its ranks, and at once entered the forest and resumed the old ground.[15] Fourteen or fifteen rebels who belonged to North Carolina concealed themselves between logs, to be sheltered

---

12 1st Massachusetts was designated as skirmishers.

13 Just as this operation was beginning Prince "received orders to cease all operations, the reason being assigned that this was the wrong road." Prince rode back to French's headquarters and explained the situation. French reversed himself and said to proceed with the advance, and call on Carr's 3/III division for support if needed. Prince returned to his command: "but finding the enemy outflanking me, I sought General Carr and informed him that Major-General French had directed me to call on him for support if I needed it, and that I wished him to go into line of battle on my left. He declined to do so. I replied 'But I order it.' He begged me to understand that it was with no personal view toward me that he declined, but that his instructions were to follow. I immediately communicated this in writing to [French], and received for answer 'The general says go on.'" [OR XXIX/1, 762-763]

14 Capt. John C. Johnston, 50th Virginia "became so enraged when his men cowered behind a ridge to escape enemy projectiles that he stretched his body as a breastworks to anyone afraid to advance his men." There were a few takers who rested their guns on the Captain's body but he was uninjured. [Mine Run, 56]

from the fire of friends and enemies during this last advance; and gladly rose up, cast aside their equipments and rifles, and, in the excess of their joy, actually threw their arms around the necks of the astonished soldiers.[16] A private who [had earlier] belonged to a detachment of Union prisoners, and a member of the [rebel] guard that accompanied them from Chancellorsville to Richmond, recognized each other, and grasped hands in the most cordial manner.[17]

"Come in!" "Come in!" the skirmishers shouted, and a number complied with the request: but one boldly yelled, "No, I don't; I'm no such man as that;" and three bullets shattered his limbs while he was trying to escape. A rebel battery opened at sunset, and continued to fire for two hours, during which the flashes lighted up the forest. The enemy abandoned the road, when quietness ruled the night.

The substitutes, with a few exceptions which are always found in a certain number of persons, bravely withstood the

---

15 The Confederate attack by Johnson's division centered on Carr (3/III). Prince wrote: "Suddenly the enemy broke forth in a tremendously noisy advancing fire, which very soon gained so much ground beyond our left that it seemed as if the enemy would turn us, and I was not much surprised when I saw my line turn and move backward slowly and reluctantly. I saw that it was general, and thereupon did not attempt to stop it till it got back where I had posted my battery." [OR XXIX/1, 764]

Johnson's attack ran into two Union artillery batteries with infantry support. Prince wrote: "the moment the battery was unmasked by our men, it opened with the utmost rapidity, deluging the rebel ranks with double charges of cannister." [Ibid.] Lt. J. K. Bucklyn, Battery E, 1st Rhode Island, reported: "About 4 o'clock the enemy charged our lines with such impetuosity that he quickly drove our infantry from the wood. With the peculiar rebel yell they came on until they reached the edge of the woods, about 30 yards in my front, when I gave them canister shot, spherical cases, and shell as fast as I could load and fire." [Ibid. 794]

16 1st and 3rd North Carolina (Steuart's brigade) were heavily engaged in the attack by Johnson's division.

17 The incident of the private and the guard apparently refers back six months when the private had been captured at Chancellorsville – Blake is calling

shock of the battle; and it is a strange circumstance that the list of killed in the regiment consisted of the original members.[18] When the brigade charged across the field near the house, some frightened hens left their nests in the bushes, and flew before the men with a shrill cackling which afforded much amusement in the midst of the hissing balls. A captain of the skirmishers basely deserted his post, and attempted to pass through one of the companies, but was halted by a faithful sergeant, who seized him by the collar, and said, "You are setting a pretty example to your men: go back, you cowardly scoundrel!" and compelled him to return to the front. The rebels in one onset rushed with yells, which the first volley converted into groans; and a soldier remarked as he elevated his piece, "If I am hit, I hope it will be a finisher;" and within the space of a minute a bullet penetrated his eye, and his wish was gratified.

The corps suffered a loss of five hundred killed, wounded, and missing, in the valueless engagement:[19] the unpardonable delay was hourly prolonged; and an almost impassable chasm still existed between the right wing and the Army of the Potomac. The ignominious result of this conflict produced intense dissatisfaction among the soldiers, who sadly pictured the brilliant victory that might have been gained if a sober and intelligent officer had wielded the immense power which had been so heedlessly conferred upon a besotted major-general.[20]

---

attention to the coincidence of their chance reunion at Mine Run.

18 The 11th Massachusetts had four men killed in action that day, all of whom were part of the original muster of June 13, 1861.

19 A later tally showed 943 casualties in III corps, with 399 in Carr's division. [OR XXIX/1, 680-682] The estimate for Johnson's division was 498. [Ibid. 837]

20 French was not entirely to blame. It was Halleck and Meade who appointed mediocre officers to lead III corps, Birney being the exception. Why Birney's division did not lead the advance is unclear – perhaps he

The happy North Carolinians who were captured by the regiment most willingly and truthfully said that Johnson's division of Ewell's corps comprised all the troops that were posted in the Raccoon-Ford Road; and the thin line of skirmishers won a part of this position upon the right, which was entirely undefended. Six Union divisions confronted one composed of rebels: a force could easily turn their left flank; and they would have been routed by the overwhelming masses of the third and sixth corps, which were aligned near this point. A division that was termed by the veterans "pets," or "lambs," because a corps general bestowed every favor upon it, was placed within range of hostile cannon for the first time; but it was demoralized by the pernicious example of this drunkard and the cowardice of its commander, and tarnished its history with disgrace, which subsequent service under brave leaders removed.[21]

The sixth corps marched at midnight in the proper direction to Robertson's Tavern: the division retired from its position at daybreak upon the 28th, and plodded slowly though the rain, which did not cease until noon, and covered the roads with mortar-beds of red mud. The corps advanced several miles upon the broad turnpike that passed through Orange Court House,[22]

---

was too closely associated with the old III corps, which had never been warmly appreciated by either Meade or French.

The staff work of Meade and Humphreys can also be questioned. For the river crossing of two corps, they picked an unfamiliar ford which had a number of deficiencies, and chose an indirect route to the army's rendezvous point near Robertson's Tavern. They provided an insufficient number of pontoons, and did not supply local guides or useful maps.

Lastly, General Prince caved in to French, against his better judgment, that the left fork at Widow Morris' was the correct route. Unfortunately, this was Prince's first divisional command – his prewar experience was largely as a paymaster.

21 Carr's "lambs" eventually became the 3/VI division, and fought well under Gen. Ricketts at the battle of Monocacy in July 1864.

22 Orange Court House is about fifteen miles west of Mine Run. Blake is

and then moved in the rear of the army from the right to the left; and the men ascertained the lines that defined the front by listening to the intermittent volleys of the skirmishers. The enemy gradually fell back; and shelters of small stones which the sharpshooters had occupied were scattered at short intervals in many places. Near one of them I saw the body of a dead rebel, who carried a haversack which was filled with his rations, that consisted of nothing except dry corn. The column bivouacked after sunset, and furnished details for picket-duty, who were ordered to report to one of that large class of staff-officers that are always inefficient in the presence of danger. Those [picket-details] that belonged to the regiment marched two miles in the night, forded runs, leaped walls and fences, and discovered with amazement upon the following morning that their companies were only a few paces in the rear of their posts.

The army had finally concentrated; and the soldiers arose at half-past four, A.M., upon the 29th, and prepared for action, not against a divided and surprised force, but one which was on the alert, and strong. The dark clouds lowered constantly during the day, and occasionally parted to remind those upon the earth of their existence; while the division upon the left of the corps made a reconnoissance, and connected in the afternoon with a large flanking force under the command of Gen. Warren. "Where are we?" some of them asked an old inhabitant.

"Ringe County," he replied.[23]

The trees were judiciously cut to guide the moving lines; but the thickly wooded country and the state of the impassable roads continually hindered them, and the powers of darkness prohibited an attack, when three signal guns were fired at half-past five,

____

being overly verbose in saying that part of their march followed the Orange Turnpike on November 28.

23 That is: "Orange" without pronouncing the "O"

P.M. A horse, laden with turkeys and chickens for the use of a general, passed by the regiment a few minutes after orders were read to the men to live upon half-rations, and became the innocent subject of many emphatic phrases. The division bivouacked at nine, P.M., and thousands of cold and weary forms clustered around the low fires which were allowed; and commands were frequently issued, "That fire is too high," "Take off that log at once," when a desire to receive comfort triumphed over caution.

November 30, 1863, is a day that will be long remembered by the troops that were massed at Mine Run; not because a battle was fought, but for the singular reason that no conflict took place.[24] The division was under arms at one, A.M.; instructions were repeated in whispers by the officers; and no conversation or unnecessary noise was permitted as it marched upon the famous plank road, from which it debouched to the left, formed in three lines, and the brigade was placed in the front and supported by the "Excelsior" and Jersey brigades. Gen. Warren commanded six divisions, comprising twenty-eight thousand men, and extending three miles, that were aligned in a similar manner for the purpose of storming the breastworks of the rebels, who had fortified the strong ridges west of Mine Run. This is an insignificant tributary of the Rapidan, that varies in depth from

---

24 On November 29, Warren moved on the Confederate right flank with the II corps and a division from the VI corps "so as to threaten it, and endeavor to discover a vulnerable point of attack, and if necessary to continue the movement, threatening to turn his right." The other corps commanders were to look for weaknesses in the Confederate defenses. On the night of the 29th Meade met with his corps commanders and decided to attack along the entire line. However, French said it would be "impracticable" in his center section. Therefore Meade decided that the VI and V corps would attack on the right and that Warren with the II corps, along with the divisions of Prince and Carr of III corps, would attack on the left. [Humphreys GR, 64-65]

three to five feet, and is crossed by the plank road near old Verdiersville. The mathematicians in the ranks amused themselves by multiplying three miles by three lines, and obtaining a product which they termed a "nine-mile charge." All the dispositions for the assault had been made before daybreak; many knapsacks and haversacks had been unslung to relieve the bearers of the weight; muskets had been stacked; and at eight, A.M., the signal-gun would resound through the forest from the right; the trusty bayonets would be fixed; the vast columns with their tattered flags would rush forward with hurrahs; and the [percussion] caps had been removed from the pieces to prevent those that carried them from discharging a bullet. The night had been excessively cold; the blankets and clothing were covered with frost; the water in the canteens was condensed into a cake of ice; and it is a sad fact that Union soldiers were benumbed and died upon the picket-posts. The fires were extinguished, and every one sought physical warmth and excitement by keeping all the limbs in motion; and squads were running, wrestling, or striking hands, for several hours; and some climbed up the trees to gather persimmons. Several unlucky pigs which wandered from the safe woods at this opportune moment were relentlessly pursued until rations of fresh pork were secured.

The division awaited the final word of command behind a hill which protected it; and most of the ground in its front was cleared, so that groups examined the works of the enemy, which had been constructed upon another elevation that rose at the distance of a quarter of a mile, and was parallel with it; and Mine Run flowed through the valley between the heights. The rebels were strengthening their positions every moment by untiring labor, which the necessity for bodily exertion greatly increased; their lines were defined by the strokes of axes and the crash of falling trees; while squads were carrying logs or plying the spade. No shells were fired: the skirmishers thrust their bayonets into the hard and frozen soil, and ran to and fro to conquer the cold, and never molested the thousands [of rebels] who were scanning their rifle-pits and redoubts. Mounted [rebel] officers posted the re-enforcements of infantry and artillery which arrived behind the walls upon which the rags of treason were conspicuously displayed; and many waved the staffs, beckoned with their hands or threw their caps into the air, and shouted, in tones of defiance, "Come on!" The spectators, including generals and privates, concurred in the same opinion regarding the undertaking; and the heart of the bravest sank within him as he gazed upon the scene, because a disastrous repulse was the certain result; and all calculated the number of steps which could be taken before the fatal bullet struck a vital part. I never beheld such a universal expression of gloom and dismay: watches and other valuables were deposited with chaplains, quartermasters, and other non-combatants; and brief epistles were written by those who felt like persons upon the couch of death. The ignoble poltroon skulked to the rear when unobserved; but those that remained resolved that they would not turn back, but advance, unmindful of canister or balls, until their feet touched the frowning crest. The soldiers watched with impatient eyes the sun, which seemed to stand still; but the long-dreaded hour came, and an active cannonade opened upon the right at eight, A.M. The ranks were promptly formed; the names of absentees were recorded; and all expected to hear the decisive order, "Forward in line, guide right!"

Gen. Warren, with a sense of honor which cannot be too highly praised, declined to sacrifice the lives that had been placed in his charge. Gen. Meade decided most wisely to abandon the proposed movement; and the position of the troops was undisturbed until night, when the elated division joined the corps.

The expediency of attacking the enemy at Williamsport [after Gettysburg] and Mine Run is determined by examining facts that clash in every respect; and it does not appear strange, when the motives are understood, that the generals who opposed the first favored the second proposition, and vice versa. When Lee trembled upon the banks of the Potomac, destitute of ammunition, disheartened by defeat and heavy losses, — surrounded by a loyal people, and victory was sure, they described his impregnable position, the unbroken morale of his army, and the insecurity of Washington if a defeat was sustained. When the circumstances were reversed, and the rebel chieftain stood in his elaborate field-works at Mine Run, and invited an assault, furnished with the supplies of war, his line of retreat open, in the midst of friends and abetters, and disaster was certain, — these obtuse military judges declared in favor of an onset by the Union forces.[25]

December 1 was unmarked by any conflict; and the brigade was detached for special duty in the afternoon, to report to Gen. Gregg,[26] to support the cavalry at Parker's Store, where the rebels had made a desperate effort to gain the road upon the preceding day. The soldiers burned the barn which belonged to a woman who gave information to the enemy concerning the number and position of the Union troops; and a strong guard was posted upon her house to save it from the same destruction. The bodies of two men (who had been killed in the action) were found near this position; and they had been entirely stripped, and left in the

---

25 This convoluted polemic is directed primarily at Meade, but also at Sedgwick, who apparently was willing to go ahead with the assault at Mine Run.

26 Brig. Gen. David Gregg, commander of 2nd Cavalry division.

woods by Stuart's cavalry. Large fires were built by the reserves to deceive the watchful line of sentinels; a volley was delivered as a signal at two, A.M., of the 2d; the pickets withdrew from the front; and the main portion of the army marched by the bivouac during the night. The brigade moved at daybreak; and this infantry force actually covered the retreat of the cavalry, crossed the pontoon bridge, that consisted of eight boats, at Culpeper "Gold-mine" Ford, and occupied the old camps at Brandy Station.[27]

------

27 "But for the restrictions imposed on General Meade from Washington, he would have fallen back toward Fredericksburg, taking up a position in front of that town. Had he done so, the first battle with Lee, in May, 1864, would not have been fought in the Wilderness, but in a more open country." [Humphreys GR, 68]

Chapter XVI

Winter Quarters at Brandy Station

THE ARMY steadily advanced in successive years from river to river, and erected its winter quarters upon the banks of the Potomac [1861], the Rappahannock [1862], and the Rapidan [1863]. The headquarters were established at the same point that had been occupied by Lee, and the [flag-]staff which he left in his hasty flight was unadorned; while the American flag daily ascended and descended the high pole when the call "to the color" was sounded at sunrise and sunset. The telegraph-office in the town was occupied by the same operator for the fifth time in the various changes that had taken place in the position of the army: the rebels always possessed it for a similar purpose as soon as it was abandoned; and both parties used the same table, and several miles of the same wire. Operations against the enemy, and drills, were suspended during the inclement season; and details to guard the trains, the camps, and the picket-lines, and labor upon the roads, comprised the routine of duty. Courts-martial assembled frequently to determine the nature and punishment of military crimes; and one tribunal, of which the author was judge-advocate, tried about forty men for misconduct in skulking from Mine Run; and a chaplain was found guilty of stealing a horse, and dismissed from the service by order of the President.[1]

The face of the country soon assumed the barren aspect of Falmouth; and the pickets of the brigade, for a month, made their

---

1 The dismissed chaplain was the 11th's Elisha F. Watson who, as Blake noted earlier, had many talents but none as a minister. The regiment apparently had no spiritual guidance after Watson's dismissal on February 24, 1864.

fires of the woodwork of corn-shelling, threshing, and the numerous machines with which a large farm was supplied; and iron rods, bolts, ploughshares, cranks, and cog-wheels were sprinkled upon the ground in the vicinity of the posts. The fifteen hundred inhabitants that lived in Culpeper before the Rebellion had been reduced to only eighty persons, who were chiefly dependent upon the Government for the means of sustenance. The court-house and slave-pen had been gutted, and were used as places of confinement for rebel prisoners. The fences that enclosed the cemeteries which were attached to the churches had been torn down and burned; and sinks, booths, stables for horses, and the fires of the cooks, were scattered in the midst of the gravestones and tombs. The state of destitution that prevailed may be illustrated more clearly by quoting the remark of a young woman who resided in the place: "My father was worth $300,000; but all his people, except a small boy, ran away with your folks; his large house was burned by your cavalry; we eat your pork and bread; and, just think of it! I haven't had a new dress or bonnet since the war began." The refugees and their families constantly entered the lines; and one of them said that he was assisted by a friend, who gave him his horse, and manifested much indignation, and declared that the animal had been stolen, to mislead the neighbors, when he received the news of his successful escape. Deserters exhausted their ingenuity in finding ways to reach the cavalry vedettes; and some gladly swam across the Rappahannock in the coldest nights of the year.

The old residents asserted that the ground upon which the division had encamped was always submerged in winter, and it would be impossible for the men to remain there until spring: but the barracks were never swept away by any inundation; and they explained the matter by saying that it was the driest season that had existed for thirty years. The results of one severe rain, that deluged the plain, showed that, if they were often repeated, all persons would perceive the wisdom of the warning. The river rose and overflowed the swamp so suddenly, that the members of seven posts which were located near it were obliged to climb trees to avoid the unlooked for danger of drowning; and the brief tour of picket-duty was extended many hours.[2] Squads that were not stationed in the forest found themselves upon an island, and waded through deep water a long distance; and some were compelled to swim to reach the reserve upon what was the main land. A small stream was enlarged to the dimensions of a lake, one-fourth of a mile in width; and a part of the cavalry provost-camp was submerged, and an officer discovered that the rushing water was two feet deep in his tent when he awoke. The weather-wisers always glanced at the mountains; and the voices of experience uttered the following precept, — that there would be rain once in every two days as long as the snow crowned the crests of the Blue Ridge.[3]

During this period the enemy did not attempt to make any movement, although a long line of railroad conveyed [Union] supplies from Alexandria; and the troops of Lee labored unceasingly, and constructed miles of earthworks upon the bluffs that had been fortified by nature; while the Union forces rested in their camps, and relied for defence upon the strong arm and loyal heart. A number of false alarms occurred, and the soldiers were sometimes ordered to be in readiness to march at a second's notice to resist an advance; and, upon two occasions, the main

---

2   The picket line of 2/III division was about three miles south of Brandy Station just in front of Stony Point, where it joined the pickets of the Second Corps. At this point it was about a mile and a half due north of Morton's Ford. The picket line then ran about three miles due west where it joined the line of I corps, just to the south of Pony Mountain. [Hall]

3   Mountain Run appears to have been in close proximity to the division's camp and may have played a role in the flooding.

body of the infantry co-operated with the cavalry, and made feints to cover the blow that was aimed at other points. At four, A.M., upon February 6, the troops were commanded to procure the usual amount of rations and ammunition for a campaign, and concentrated near the fords for the ostensible purpose of crossing. The division halted in a swamp, about a mile from the river, and acted as a support for the second corps; one division of which forded the stream, gained a position upon the opposite bank, and recrossed at night. No shots were fired on the 7th; and the army, having attracted the gaze of the foe, withdrew late in the afternoon, at the time that an expedition was in motion upon the Peninsula against Richmond.[4] Another demonstration was made upon Madison Court House on the 28th to conceal the raid of Gen. Kilpatrick.[5]

The number of officers' wives and other ladies that were present in the camps was much larger than at any previous period; and balls and similar festivities relieved the monotony of many winter quarters.[6] Large details, that sometimes comprised a thousand men, were ordered to report at a certain headquarters for the purpose of constructing suitable halls of logs and the "sacred soil" of Virginia. A chapel was built within the limits of

---

4 In the first week of February Gen. Benjamin Butler wished to launch a major cavalry raid on Richmond from Williamsburg and asked the Army of the Potomac to provide a diversion on the Rapidan. II corps was to cross Morton's Ford with the 2/III division in support. They crossed the river and engaged a portion of Ewell's Corps until sunset, when they were withdrawn.

5 Gen. Sedgwick moved on Madison Court House on February 28 while Kilpatrick's 4,000 cavalrymen raided Richmond.

6 General Carr hosted a III corps officers' ball at his elaborate headquarters at "Sunbright" mansion in late January. II corps, not to be outdone, put on an even larger affair on Washington's Birthday. For a fuller description of Gen. Carr's ball and the officers' social life, with particular emphasis on Gen. Humphreys, see Mutiny, 33-38.

the brigade by soldiers, who daily labored upon it for three weeks; and many of the officers contributed money to purchase whatever appeared to be required for it. An agent of the Christian Commission furnished a capacious tent[7] which formed the roof; and religious, temperance, and Masonic meetings were frequently held, until this apostle, who employed most of his time in writing long letters for the press, that portrayed in vivid colors the "good work" which he was accomplishing, removed the canvas because an innocent social assembly occupied it during one evening. The enlisted men, who rarely enjoyed the benefit of these structures which they erected, originated dances of a singular character.[8] By searching the cabins and houses of the natives, and borrowing apparel, and a liberal use of pieces of shelter tent and the hoops of barrels, one-half of the soldiers were arrayed as women, and filled the places of the seemingly indispensable partners of the gentler sex. The resemblance in the features of some of these persons was so perfect, that a stranger

---

7   The use of the chapel tent for social activities exacerbated the conflict between Colonels Blaisdell and McAllister (11th New Jersey) which had been growing since the late summer of 1863 as to whom would head the 1st Brigade. McAllister believed that Blaisdell was lowering the moral tone of the brigade and undermining his work in getting his regiment to regularly attend church. McAllister "was known as a sincere Christian, but of rather a puritanical bent, and strongly opposed to the use of intoxicants in any form. Blaisdell, on the other hand, was not particularly known as a teetotaler, and did not object to the men having their regular ration of stimulant. Though both colonels were brave and gallant, they could not, because of disparity of taste and dispositions, become very warm friends." [Marbaker, 152]

8   McAllister wrote that Blaisdell was "the party that was prominent and conspicuous in taking the church for a dance hall." He noted further, on learning of the enlisted man's dance: "some of our chaplains called on Gen. Prince, stated the facts to him, and desired that it be stopped, but Prince decided against the wishes of our chaplains, saying at the same time that he had often attended balls in churches. (It happens that he is a Unitarian.) … The Chaplain of the 11th Massachusetts is a poor, miserable nobody, or he might have prevented it. [McAllister, 393]

would be unable to distinguish between the assumed and the genuine characters.

The sewers of recruits and substitutes that had received enormous bounties, and possessed the same qualities as their predecessors who joined the regiment at Beverly Ford,[9] continually flowed into, or, to speak with more accuracy, through the army. Those that had served in the rebel ranks were sent to the north-west to assist Gen. Pope in subduing the Indians; others who had been seamen floated into the navy; and some entered the hospital: so that, when the spring campaign opened, only a small fraction crossed the Rapidan. A number that arrived at the camp after sunset escaped before morning; so that the officers of the company who had not seen them could not identify them when they were arraigned and tried, for desertion, before a court-martial. The thousands of crows rendered more actual service than a majority of this class of persons, and devoured the entrails of animals which had been slaughtered by the butchers, and the carcasses of dead horses and mules. They were never shot, because the citizens had no guns, and the soldiers would be punished if they wasted ammunition; and they grew tame and fat in opposition to the well-known saying,[10] and propagated so rapidly that their immense numbers blackened acres of ground in the vicinity of the camps. One noticeable event was a fire which swept over the field of Cedar Mountain, and caused the explosion of shells that had remained there nearly two years after the battle.

An episode occurred in my military career, which may not be of general interest; but an omission to allude to it might

---

9　The previous intake of substitutes had been on August 23, 1863 – mentioned by Blake in Chapter XIII.

10　We were unable to identify a "well-known" saying about crows relating to their growing tame and fat.

produce a slight degree of surprise and criticism. I was detailed to act as judge-advocate of a court-martial in the morning, placed under arrest in the afternoon, and transported with four officers in a wagon which was marked very conspicuously "Provost Guard," and followed by a detachment of soldiers to the headquarters of the corps. I was closely confined three weeks in a log shelter in which there were no windows, but the rents in the roof admitted light. No friends were allowed to visit the quarters unless the corps commander granted permission; and sentinels constantly paced their beats at all hours, and watched the prison, because I had acted as secretary of an orderly meeting of officers that adopted resolutions of the highest loyalty and patriotism, which were duly transmitted to Gen. Meade and the Chairman of the War Committee of Congress. The perjury of three unscrupulous witnesses complicated the case; and, while some were dismissed from the service, a heavy fine was imposed upon the author, and subsequently remitted by the commander of the corps, who was convinced of its utter injustice.[11] A field-officer

---

11 Blake's court-martial grew out of Gen. Meade's controversial army
   reorganization, March 25, 1864. The I and III corps were abolished, and
   their regiments assigned to other corps. Several junior officers of the old
   Hooker Brigade wrote what in retrospect seems a very mild criticism.
   However, the leadership of the army took it more seriously. Along with
   Blake, the other defendants were Capt. James R. Bigelow, Capt. Walter
   Smith, Lt. George Forrest, (11th Massachusetts), and Capt. E. C. Thomas
   (26th Pennsylvania). Col. McAllister saw the 11th Massachusetts
   defendants as part of the Blaisdell "clique" and that his only regret was that
   the colonel himself was not arrested. The charges were filed by General
   Carr, declaring that the defendants actions in preparing and passing the
   protest resolution tended to "incite a mutiny." One of the defendants, Capt.
   Bigelow, had also called Gen. Carr a "cowardly son-of-a-bitch," in a
   private conversation prior to the meeting, thus he was also charged with
   "contemptuous and disrespectful conduct" to a superior officer. The
   principal prosecution witnesses were all from the 11th New Jersey. For a
   full discussion of the army reorganization and the courts martial from the
   original transcripts of the trial the reader is referred to *The Mutiny at
   Brandy Station.*

of a regiment, who enforced the principle established by this decision and sentence, dispersed an assembly of subalterns that had convened for the purpose of taking measures to send home the remains of a comrade who was killed at Mine Run.

The ordinary preparations for active operations were made as soon as the roads became dry and hard: the ladies were notified to leave the camps previous to a specified date; surplus baggage resumed its annual visit to the storehouses in the rear; and reviews, inspections, and target-practice, daily took place. The army was re-organized, the troops were consolidated into three corps, and the [second] division which had always belonged to the third [corps] constituted the fourth division of the second corps. The ["Hooker"] brigade, which was one of the oldest in the volunteer service, was dismembered; and the needless separation of the regiments that had so long fought side by side the common enemy resembled the breaking-up of a family. Gen. Meade displayed a censurable ignorance, or lack of judgment, when he assigned certain generals to their positions; and one division was almost demoralized by the appointment of a notorious coward, knave, and ignoramus to the command.[12] A general, who always deserted his brigade whenever the trials of battle demanded his presence; who never discharged the numberless accounts of the sutlers and commissaries for the food, rations, and liquors which he consumed; who employed escaped negroes as servants, and defrauded them of their just compensation; who displayed a profound ignorance upon every subject, which made him the butt of ridicule for soldiers of all

---

12 Here Blake embarks on his most strident criticisms of Joseph Carr. Considering his military record alone there is no reason to suspect Carr of cowardice. He led the 2nd New York during 1861 and 1862 in the battles of Big Bethel, Peach Orchard, Glendale, Kettle Run, and Second Bull Run, and was ultimately promoted to divisional command. The only positive trait Blake noticed in Carr was that he was a "good drill master."

ranks, from the highest to the lowest; who had been originally commissioned as colonel by the influence of a base pugilist of New York, who was promoted because he was a foreigner,[13] — was crowned with honors when he merited disgrace. Gen. Alexander Hays of Pennsylvania, a graduate of West Point, one of the most fearless and honorable officers in the national forces, whose pre-eminent gallantry infused confidence upon doubtful fields, was degraded from the command of a division to elevate this Celtic vagabond.[14] A meritorious commander, who bore upon his person the scars which are the soldier's badges of honor, annoyed the headquarters of the army by preferring charges of cowardice against one of these favorite division appointees, and was ordered to report for duty in the West.[15] The words and deeds of Gen. Hancock were narrowly scrutinized, to detect, if possible, some pretext upon which to base his removal; and his official action was continually embarrassed by this contemptible surveillance.[16] Fortunately, Congress revived the grade of lieutenant-general, and the soldiers rejoiced that the pernicious

---

13 Carr was born in Troy, NY of Irish parents. He was instrumental in raising the 2nd New York in 1861, but command was instead offered to George L. Willard of the 19th U.S Infantry. When Willard declined the offer, the state turned to Carr. We have not identified a candidate for the "base pugilist of New York."

14 Hays lost his division, but commanded a brigade under Birney which was larger than his previous command. Hays was killed in action May 5, 1864.

15 Prince raised competency and bravery issues as to Gen. Carr at Mine Run. Ironically Prince had been given command of Carr's "lambs," (3/III), which in the reorganization had been redesignated as 3/VI. Prince was sent to the District of Kentucky, on April 28, 1864, and replaced by Gen. James Ricketts. [Callum]

16 Carr's designation as commander of the 4/II division (comprising the remnants of the old III corps), was successfully challenged and Carr was replaced by Gen. Gershom Mott, of the Jersey Brigade. Hancock was just back from sick leave on account of his Gettysburg wound and felt somewhat uneasy in taking a strong position on Carr's appointment which was the work of Humphreys and Meade.

influence of Gen. Meade was seriously crippled. All anticipated the removal of this officer, — a question that had been often discussed in the newspapers and around the camp-fires; but the wise conduct of Gen. Grant, who, unlike his immediate predecessor, Gen. Halleck, preferred the dangers of the front to the safety of the rear, fixed his headquarters, not at Washington, but at Culpeper Court House, and thereby saved the Army of the Potomac from the annihilation which awaited it under incompetent commanders.

Chapter XVII

The Wilderness and Spottsylvania[1]

THE FIFTH CORPS, which guarded the line of communications – the Orange and Alexandria Railroad – during the winter, and the ninth corps, joined the main army upon the first and second days of May: the log huts were destroyed, and the troops bivouacked in the fields, to prevent any delay in marching whenever the orders were received. After sunset, [May 3] the commands were issued for the regiments to move at half-past ten, P.M.: all unusual fires were prohibited; the tattoo and taps at the stated hours were heard by many for the last time; and the long columns and trains advanced to the fords of the Rapidan at one, A.M.,[2] upon [May] 4th. The veterans, that had made their foot-prints upon many lengthy and dreary roads, reserved their strength, and silently followed the file-leaders; and verdant recruits and substitutes were the only babblers. The beginning of this campaign was like all those which had preceded it; and thousands of overcoats and blankets were scattered in the woods and fields through which the soldiers passed.[3] The cavalry gained the commanding heights of the Rapidan without a contest; and the regiment crossed the river at Ely's Ford at half-past twelve, P.M., upon a pontoon bridge that consisted of nine boats which

---

1 The name, as it relates to Virginia, derives from the 18th-century Governor of Virginia, Alexander Spotswood., and "sylvania" is Latin for "woods."

2 The original text says "one, P.M." but Hancock's II corps set off just after midnight heading for Ely's Ford on the Rapidan. The head of the column arrived at Chancellorsville by 10 A.M. May 4th. [Humphreys VC, 18-19]

3 The Army of the Potomac's habit of throwing away overcoats on the first day of a march did not favorably impress their new commander. Grant wrote that it was an "improvidence I had never witnessed before." [Grant/2, 19]

were anchored twenty-one feet apart. The felled trees and other obstructions in the roads were removed by the axes and spades of the pioneers; and the troops marched with great rapidity until three, P.M., when the corps halted, and formed its lines upon the old battle-ground of Chancellorsville. The numerous breastworks that were thrown up by both armies to hold their positions, the shattered oaks and splintered limbs, and the fragments of weather-stained clothing and equipments scattered upon the field, reminded the men of the familiar scenes of that sanguinary struggle. Many of the Union dead had been exhumed, or remained unburied; jaws, arms, and legs were bleaching upon the soil; and the wasps and moles made their nests in some of the skulls. Not a shot was fired during the day; no bugle or drum resounded through the forest; the unnatural stillness which precedes the dreadful tempest reigned; and the brigade bivouacked upon the same spot that it occupied on the same date of the previous year.

The soldiers were awakened at three, A.M., on [May] 5th; the humble meal termed a breakfast was prepared; and they marched at five, A.M., to the ruins of the Chancellor House, from which the corps under the command of Gen. Hancock pushed forward towards Spottsylvania Court House.[4] The hordes of Lee emerged from the fortifications which had been rendered useless by the passage of the Rapidan;[5] a halt took place near Todd's

---

4   The II corps was headed toward Shady Grove Church, south of the Wilderness. It was then supposed to link up with Gen. Warren, now commanding the V corps, who was guiding his troops towards Parker's Store on the Orange Plank Road. Gen. Lee, however, did not do what Meade assumed he would — a defensive stand as at Mine Run — but rapidly advanced to attack up both the Plank Road and the Turnpike. The Union cavalry under Gen. James H. Wilson failed to keep Meade informed as to the whereabouts of Lee's forces.

5   Rendered useless because they had been flanked by the Union move.

Tavern at ten, A.M.; and the corps was ordered to change its direction, and hold the junction of the Fredericksburg Plank Road and that which ran from Germanna Ford.[6] The report of a rifle occasionally indicated the presence of the enemy; and at twelve, M., the first cannon sounded the prelude to the battle of the Wilderness. This chapter, it may be needless to remark, does not narrate the movements of the entire army, because the character of the country, and position of the author with the rank and file, limited his view; and the incidents of a part of the lines, that extended five miles, are described. The division hastened to the vital point, which it reached in the afternoon, and remained in the reserve,[7] rushing from post to post, until 4.10, P.M., when the musketry in front assumed the prolonged roll that always marks a heavy engagement. There were no commanding elevations or open tracts of ground upon which the artillery could be planted; and the firing was necessarily restricted to the small-arms, that slay the tens of thousands while the shells and solid shot destroy

---

6　Meade learned of the advancing Confederate columns in the early morning and abandoned the original plan in favor of an immediate advance down the two roads to meet the attackers. He ordered II corps, on the Brock Road, to countermarch to the Plank Road intersection, the objective of the A. P. Hill's Confederate corps. When the about-face came, II corps was spread out along six miles of a narrow road. Communications between Hancock and Meade were unaccountably slow, and Meade operated on an equally unexplainable assumption that Hancock would be in position to launch an attack up the Plank Road in the early afternoon.

7　"It is no small matter to bring up twenty-five to thirty thousand men by a single road and form them for battle; and the difficulty was, in the present case, increased by the narrowness of Brock Road, and the density of the woods on either side. The greatest efforts of the staff were put forth to hasten the work; to get the artillery out of the way and to push the infantry forward." [Walker, 414] Getty's 2/VI division had been rushed forward to defend the Brock Road - Plank Road intersection about noon, barely beating Heth's division of A. P. Hill's Corps. Birney's division (3/II), now leading the corps, arrived at about 1:00 P.M. and Mott arrived at 2:00.

the thousands. The great contest[8] occurred in the midst of an almost impenetrable jungle of scrub oak, decayed trees, dense underbrush, and short pines, in which a regiment could not be discerned at the distance of a hundred feet; and the proper formation of the ranks seemed to be an impossibility. It was rarely intersected by public ways over which the infantry could move; and the pioneers were continually engaged in felling trees and cutting new roads to facilitate the communications between the right and the left. Packed mules, which transported axes and shovels, were attached to every brigade, and formed an invaluable auxiliary during the campaign. Breastworks were hurriedly constructed to defend the Germanna-Ford [Brock] Road; and the dry logs of which they were principally composed were easily set on fire, so that it was often necessary to remove a part to save the rest. Sunset came: the darkness of the night followed, but did not check the din of the conflict, which continued when the combatants were unable to perceive friends or enemies, and suddenly ceased at eight, P.M.[9] The Union forces

---

8   The exact details of the "great contest" are open to conjecture. Excelsior Brigade commander, Col. William Brewster, never prepared a report because he left on sick leave May 5 and did not report back for duty until May 28. Col. Blaisdell, who assumed command of the brigade, was killed within a month and never had an opportunity to write a report.

9   Precise details of the day's fighting depend on whose account is consulted. Recent portrayals of the Wilderness (Rhea, Priest, and Scott) have relied on Col. Robert McAllister, commanding the 1/4/II brigade, who wrote that the right end of his brigade came in behind Getty's left. As it advanced, Mott's division apparently drifted to the right toward Birney's division that had moved to support Getty astride the Plank Road. McAllister noted the crowding in the center of his line and declared that it was impossible, in the dense underbrush, to form in two lines.

    The Excelsior Brigade, on McAllister's left, apparently was able to form in two lines and even entrench to a degree. Lt. Col. Michael Burns, 73rd New York, reported: "The works were scarcely constructed when the first line of battle was fiercely assailed by the enemy. ... The first line having broken fell back and this command was soon attacked, and after sustaining the shock for fifteen or twenty minutes, ... the Brigade was

did not yield a single position; and every attempt which was made by Lee to overpower the columns that were advancing by the flank was successfully baffled.[10]

The wounded and dying were borne upon stretchers to the hospitals in the rear; and the usual number of skulkers sought to escape the perils of the battle by travelling in the same direction, and eluding by ingenious devices and shams the vigilance of the provost-guard. The colonel [Blaisdell] halted this class of persons whenever they passed through the regiment, and detained them if he was satisfied that they were neglecting their duties. Many conversations like the following ensued between the colonel and the members of these squads; and the questions and answers show clearly the rank and intentions of the parties, without any explanation: —

"My good man, where are you going?"

"I'm sick, and the captain told me to go to the hospital."

"Have you got a pass?"

---

obliged to fall back to the breastworks which had been thrown up by the first line, which was done in good order. The enemy followed up his temporary success, but on coming within range of the breastworks they were most signally repulsed." [OR XXXVI/1, 503-504]

McAllister stated that the Excelsior Brigade was the first to fall back, followed by the 1st Massachusetts and the rest of his brigade. However, Burns' report is cited by Steere in questioning whether the onus of failure should rest wholly on the Excelsiors. [Steere, 214-215]

The historian of the 1st Massachusetts, McAllister's leftmost regiment, attributed the retreat to encountering sudden and overwhelming firepower: "The men went forward, however, in very irregular lines, and keeping as closely together as they were able. They had advanced thus only 500 or 600 yards from the road when, directly in front, the unseen enemy opened a double volley, which sent thousands of bullets crashing through the woods into their faces. ... Along the whole division the movement became at once and rapidly retrograde." [Cudworth, 459-460] Mott's 4/II division fell back to its first position on the Brock Road.

10 "...whenever the Federal troops moved forward, the Rebels appeared to have the advantage. Whenever they advanced, the advantage was transferred to us." [Cudworth, 460]

"No."

"You are weak, and find it hard to travel?"

"Yes, I can hardly walk."

"The hospital is two miles from here; and you are used up, and can't go there. Rest here with my brave men; and I will take your name, and notify your officers if you are killed or wounded."

The soldier, knowing he could not extricate himself from the toils of his pretext, usually pleaded another, which was equally shallow; and sometimes attempted to run away. The colonel at once denounced him in language which could not be strengthened in its style, and concluded by uttering his customary orders upon similar occasions.

"Captain _____, detail one of your trusty men to report to me with a loaded musket."

"Private _____, you are responsible for this cowardly skulker. If he tries to get away, blow his brains out; but, if we are fighting, crack his skull with the butt of your gun, and he will never trouble you again."

Thirteen stragglers of this description, and others who pretended to be searching for their regiments which they had lost, were collected during the afternoon by this summary process, and assigned to different companies in the regiment.

There was little picket-firing in the night; and the sleep of those who were not upon the outposts was undisturbed until 4½, A.M., of [May] 6th, when the divisions were massed for a renewal of the battle;[11] and the musketry recommenced in the depth of the vast forest at 5.10, A.M. Squads of rebel prisoners were frequently taken to the rear, and many friendly remarks

---

11 Despite the pleadings of Gen. Heth, A. P. Hill did not order his corps to entrench after the fighting on May 5, believing that sleep for the men was more important.

were interchanged; and one of them said, "Your fellers went over our breastworks this morning like rabbits;" and, "In four rows git, and march endways." The brigade moved forward at 5.30, A.M., to support the advance, and within a brief period constituted a part of the front, and a fierce engagement followed. The men reclined upon the ground, and returned the fire of the enemy until forty rounds of cartridges were exhausted. There was a most earnest clamor for cartridges; and the boxes of the slain and wounded were opened and emptied, and a supply of those that were fitted for rifles, but unsuited to the caliber of the smooth-bore musket, was issued to the regiment in this distressing emergency by some blundering official.[12] The proper balls were brought up after a perilous delay, although some of these cartridges consisted of a solid cake of powder; and some exhibited a feeling of discontent because there were no buck-shot.[13] The bullets beat an unpleasant discord by striking the trees, which were clipped from the roots to the top, that was sixty or seventy feet above the ground.

The firing indicated at this time, when the brigade was posted half a mile in front of the Germanna-Ford [Brock] Road, the singular formation of the troops that were invisible upon the right and left. A force which was compelled to leave its position fled through the regiment, when the soldiers supposed that they were retreating to the reserve; and soon a compact mass of men was enclosed in a *cul-de-sac*, and the foe pressed closely upon the front, and left flank. They made a detour to the right, crossed the plank road, reached the original line of earthworks at mid-

---

12 It appears the 11th Massachusetts could have used their recently court-martialed Quartermaster Lt. George Forrest, who might have prevented the ammunition problems.

13 The 11th Massachusetts was still using Colonel Blaisdell's favorite smoothbore "buck and ball" muskets.

day; and the ground that had been gained by the corps was lost.[14] The extreme heat of the day increased the fatigue, and tears were shed by some who overrated the serious results of the disaster. The slaughter in many regiments had been large; and at one point the bodies of the killed remained in the places where they fell, and defined with terrible exactness the position held by the Union troops; and a long line of rebel corpses was extended in front of it. Some of the recruits, who joined their commands about forty-eight hours before the army evacuated its winter quarters, were slain in this encounter. One of the flag-staffs of the regiment was severed by a bullet, and each hand of the bearer grasped a piece of it.

The fighting, like that upon the preceding day, was confined to the infantry, on account of the impracticability of using the artillery: only three shells were thrown by the rebel gunners; and upon the Union side two brass pieces of a Maine battery swept the plank road with canister.[15] The dislodgement of the advanced [Union] force was not sharply followed by the enemy, and few bullets interrupted the rule of quietness during the succeeding

---

14 The divisions of Mott (4/II) and Birney (3/II) were attacked by Kershaw's division of Longstreet's corps at 6:00 A.M. By about 8:00 A.M. the Confederate attack was spent, but the Union forces were still greatly disorganized. Around 10:00 A.M. the fighting died down. Lt. Col. Burns wrote that "the line halted and remained in position for two hours, skirmishing going on continually. ... At the expiration of that time the enemy attacked the whole line in strong force. The regiments on our left were soon broken, and this command found itself confronted by heavy masses of the enemy on its left flank and rear. It accordingly fell back in good order to the breast-works it had held at day-light." [OR XXXVI/1, 504]

Gen. Longstreet sent four brigades down the depressed bed of the unfinished railroad [running from Orange Court House to Fredericksburg] around the Union left flank. They went down the railroad undetected, exited behind the Union line, and charged north toward the Plank Road.

15 Part of Capt. Edwin Dow's 6th Maine Battery.

four hours.[16] Squads which had been separated from their companies in the confusion attending the retreat through the bewildering thicket continually re-enforced the ranks. The division was posted once more behind the slight breastwork which had been erected upon the Germanna-Ford [Brock] Road; the skirmishers were deployed in its front at four, P.M., and the author commanded the detachment from the regiment. The groups were properly aligned within the next ten minutes, when the tramp of a heavy force resounded through the woods. Orders were excitedly repeated, "Forward!" "Guide right!" "Close up those intervals!" and finally a voice shouted, "Now, men, for the love of God and your country, forward!" The legions of Longstreet advanced without skirmishers; the veterans trained by the experience of three years beheld — *A horrid front of dreadful length.*[17] — the muskets of the feeble line were discharged to alarm the reserve; the men upon the outposts rushed to the main body; and thousands of glistening gun-barrels which were resting upon the works opened, and the fusilade began.

The soldiers crouched upon the ground; loaded their pieces with the utmost celerity; rose up, fired, and then reloaded behind the shelter; so that the loss was very slight; while the enemy suffered severely, as the trees were small in size, and there was no protection. The only artillery that was used in the afternoon was planted upon the left of the brigade, and consisted of four cannons, which hurled canister, shell, and solid shot, until their ammunition was exhausted. Unfortunately, the dry logs of which the breastwork was formed were partially covered with earth;

---

16 Like Stonewall Jackson, Gen. Longstreet was shot by his own men as he rode down the Plank Road to implement the next phase of the assault. The wounding of Longstreet caused a delay of about four hours. His corps was aligned in every direction, and Gen. Lee needed time to reorganize it.

17 John Milton, *Paradise Lost*, Book I.

and the flames, ignited by the burning wadding during the conflict, — an enemy that could not be resisted as easily as the myrmidons[18] of Longstreet, — destroyed them, and every second of time widened the breaches. The undaunted men crowded together until they formed fourteen or sixteen ranks; and those who were in the front discharged the guns which were constantly passed to them by their comrades that were in the rear and could not aim with accuracy or safety. The fire triumphed when it flashed along the entire barrier of wood, and reduced it to ashes, and forced the defenders, who had withstood to the last its intolerable heat, to retire to the rifle-pits which were a short distance in the rear. The shattered rebel columns cautiously approached the road; but the impartial flames which had caused the discomfiture of the division became an obstacle that they could not surmount. The same misfortune followed the Union forces, and no exertions could check the consuming element; and the second line was burned like the first. The conflagration in the road had nearly ceased at this time; the enemy yelled with exultation; the odious colors were distinctly seen when the smoke slowly disappeared; a general charge was made, which resulted in the capture of the original position; and the pickets were stationed half a mile in the advance at sunset without opposition.[19]

---

18 From Greek mythology; a race of men created from ants (myrmex) — followers of Achilles in the Trojan War.

19 Lt. Col. Burns wrote: "The enemy followed up its advantage, but with no better success than the day previous, he being most signally repulsed. At this time, however, the breast-works, formed as they were of dry logs and brush, caught fire and soon became untenable. The command had already expended all its ammunition. It accordingly fell back to the second line of works, and after-ward was formed as a second line of battle some 300 yards in the rear and re-supplied with ammunition." [OR XXXVI/1, 504]

The charge that pushed the Confederates from their furthest advance on the Brock Road is generally attributed to the brigades of Samuel S.

Many were eating their dinners when the assault commenced; and an officer hurriedly rushed to the works with a spoon in one hand, and a fork in the other. A panic-stricken skulker created a laugh, in which the division general [Mott] joined, by crawling upon his hands and knees from the brigade to the woods during the fighting. An enormous quantity of fixed ammunition was expended, — most of the soldiers of the division used sixty or seventy cartridges;[20] and fingers were blistered by the muskets, which became very hot in consequence of the rapid firing. The guards of both armies, in charge of prisoners, frequently lost their way in the labyrinth of stunted oaks, and entered the wrong lines, where the relations of the parties were transposed. The hurrah which has always been a characteristic of the national army was modified, and resembled the yell of the enemy to such an extent, that it was impossible to detect by this means the success of loyal or rebel charges; and some were occasionally disheartened by the joyful shouts of their friends in the distance. The division acted with the reserves, but its services were not required during the day; and an abstract of

---

Carroll (3/2/II), and John R. Brooke (4/1/II). In his report, McAllister wrote that the 11th New Jersey and 16th Massachusetts "advanced and took possession of the front works" making 'a handsome charge.'" [Ibid. 490]

Captain Edwin Dow's 6th Maine Battery, firing canister point-blank into the Confederates on the Brock Road also played a major role in the Confederates' decision to retire. Col. James R. Hagood, 1st South Carolina, wrote "Under a destructive fire I attained the enemy's works and drove him from them. He retired to a second line [held by McAllister's Brigade], keeping up a terrific fusillade, assisted by several pieces of artillery. ... My men and ammunition almost exhausted, I deemed it inexpedient to attempt anything further." [OR XXXVI/1, 1068-1069] Gen. Hancock also felt it inexpedient to launch a new attack, because of low ammunition and supply wagons which had bolted to the rear. Meade, at 5:45 P.M. somewhat reluctantly agreed. [OR XXXVI/2, 446-447]

20 The fighting in Wilderness did not follow the typical Civil War model. The lack of artillery has already been noted, and there was a considerably lower

my notes shows the slight knowledge that I possessed of the manœuvres of the army and the events that transpired. Musketry firing was often heard upon the right and left; skirmishing was active at times in the front; and columns moved in different directions,[21] and sometimes passed by each other upon the road. Lee was at last outgeneralled, and forced to abandon his position; and the regiment performed its duty upon picket without molestation; while the troops marched in the night towards Spottsylvania Court House. The companies retired at eleven, A.M., of the 8th, from their posts in the road, which was wholly deserted by the infantry; and small squads of cavalry patrolled the lines that had been so recently defended by 130,000 soldiers.[22] In the march from these scenes which now form a glorious page of the national history, some of the men were so much exhausted by the hardships that necessarily followed the great contest, that they fell asleep if they sat upon their knapsacks during a brief halt. Breastworks were built to hold the new position; rations were issued; fires were prohibited during the night; and, in the language of those whose terms of service soon expired, "there was one more day less."

The brigade was under arms at four, A.M., upon the 9th, and marched to Todd's Tavern, — the same point which the corps reached on the 5th, and upon which the cavalry fought the decisive battle that turned the rebel right and compelled the

percentage of officers among the killed and wounded because "the greater part of those who fell were struck by men who could not even see them." Only five percent of officers were killed and wounded in the Wilderness compared to eight percent at Gettysburg. [Walker, 438-439]

21 Grant was beginning his sidestep move to the left toward Spotsylvania Court House. The Fifth Corps moved first down the Brock Road on the night of May 7 while the Second Corps held in place.

22 A slight overstatement. At the start of the campaign, April 30, the Army of the Potomac had about 100,000 men, plus another 20,000 in Burnside's separate Ninth Corps.

retreat of Lee.[23] The column halted near a group of mounted officers, among whom were Gen. Grant, one of the greatest, and Gen. Meade, one of the smallest, warriors that have led an American army. The corps commanders reported at this hour for instructions; and the attentive soldiers observed, with increased confidence in the successful result of the campaign, that Gen. Meade did not give a single direction, and that Gen. Grant alone was the controlling mind. "Gen. Meade is nothing but an adjutant for Gen. Grant;" "I'm of more account with my musket than he is now;" "They don't notice him so much as they do the orderlies," — illustrate the style of the remarks that were frequently uttered by the rank and file who were interested spectators. The ambulances were insufficient for the transportation of the wounded; and the generals, with exalted philanthropy, tendered their private wagons, which were used several days for this object. An inexperienced heavy-artillery regiment,[24] numbering twenty-eight hundred men, performed picket-duty, and continually discharged volleys at the bushes and other imaginary enemies; and a sleepless night followed.

The division, guided by a negro of eminent dignity, marched at 3.40, A.M., on the 10th; halted in the road at seven, A.M., about two miles from Spottsylvania Court House; and the open fields were viewed with delight by those that recalled the horrors

---

23 Union cavalry under Sheridan fought a day-long battle with Confederate cavalry at Todd's Tavern May 7 to clear the Brock Road. Sheridan won the battle, but it delayed the Union advance several hours.

24 The Army of the Potomac experienced a severe manpower shortage as the summer of 1864 approached, due mainly to expiring enlistments for the three-year regiments of 1861. As a result, a number of heavy artillery units which had been stationed in the forts defending Washington were employed as infantry through the end of the war. Blake exaggerates the size of these units — the nominal strength of a heavy artillery regiment was 1800 men. [Fox, 5-6]

of the Wilderness.[25] The skirmishers were deployed, and drove those of the enemy until they reached the intrenchments; and the line of battle was established in an advanced position; and a belt of woods, comprising pines of large growth, intervened between the hostile armies. Labor upon the breastworks was stimulated by the exploding shells; rail-fences in the vicinity were speedily demolished; the small stones were collected; a few outbuildings were torn to pieces to make the revetment; the stakes were bound together with strips of cloth, which the men tore from their overcoats and blankets; and green boughs were placed upon the logs to protect them against the fire. The officer in charge of the packed mules remained in the rear; and the soldiers who were unable to procure shovels and axes scooped up the earth with their dippers and tin plates. The news that Johnston had been forced to evacuate Dalton was officially communicated to the troops at noon, with special instructions to avoid cheering, because the foe might ascertain the number and position of the troops.[26] The skirmish fire was incessant; the cannonading was very heavy during the afternoon; and the floating clouds of dust and smoke, three or four miles upon the left, showed the progress of the corps of Burnside. The division made an unsuccessful charge at half-past five, P.M.,[27] previous to which the officers of the regiment were told that there were probably not more than two hundred sharpshooters behind their works; but the instant

25 Mott's 4/II division was detached from the corps on May 10 to fill the gap between the VI and XI corps. As a result, it took up a position due north of the bulging "mule-shoe" salient the Confederates had erected to the north Spotsylvania Court House.

26 Joseph E. Johnston made a stand against William T. Sherman during the Atlanta Campaign at Dalton, Georgia.

27 Blake glosses over what was a significant event for Mott's division, resulting in a further diminution of its already suspect reputation due to its poor showing in the Wilderness. See Appendix: "Mott at Spotsylvania."

that the movement commenced, loud yells arose, which showed the presence of a superior force.[28]

The troops were aroused at 3.25, A.M., on the 11th, in the midst of a severe skirmish engagement; and the division was transferred to a point in the centre. A light shower of rain, which was the first unpleasant weather that had occurred since the campaign opened, fell in the morning. Musicians usually lurk in the rear; but a band that was sheltered by the line of breastworks in the front played martial airs at intervals, and invariably enlivened the soldiers, who loudly cheered.[29] The watchful sharpshooters pierced with their unerring rifles every object that might be a human being; the cannon resounded occasionally; but there was no serious battle during the day, although the sixth corps was massed at the right centre for an extensive movement, and withdrawn at sunset. The division re-occupied at midnight the earthworks which it constructed upon the 10th; and preparations were made for a grand charge by the corps,[30] the brigades of which were aligned and assigned to their positions in the course of the next three hours. Nature, that had so often favored the national cause, deployed its powerful forces; the night was darkened by the clouds, which sometimes touched the earth; no camp-fires glowed within the Union lines, while those of the enemy reflected upon the heavens like the northern lights. A dense cloud of mist, that concealed every moving body of

---

28 The "two hundred sharpshooters" actually turned out to be the 11th Massachusetts' nemesis from Mine Run, Gen. Edward Johnson's division of Ewell's Corps, along with twenty-two artillery pieces.

29 Possibly the band of the newly-arrived (and ill-fated) 1st Maine Heavy Artillery. Most three-year regiments had dispensed with their bands.

30 The positive result of Emory Upton's attack of May 10 was that Grant came to believe that a mass attack against the "Mule Shoe" salient, with less warning to the Confederates, might be successful. On May 12 the entire II corps would test this theory.

troops, filled the air at twilight; the columns received the final order to advance at 4.40, A.M., of the 12th; and thousands of hearts trembled with anxiety, as they silently and firmly approached the unknown dangers of the rebel stronghold. The [rebel] pickets, whose vigilance had been lulled by the unfavorable character of the elements, were surprised before they could awaken their comrades in the reserve, most of whom were sleeping behind a formidable earthwork, which was gained without firing a shot. While the supports were anticipating a dreadful volley, a spectacle which seemed like a dream greeted their delighted eyes. The faded banners of ungodly rebellion; two chieftains, — Johnson upon a horse, and Steuart on foot;[31] hundreds of prisoners of different grades; batteries and artillery-horses driven by the happy conquerors, — these trophies of Union success passed to the rear of the scene of action. A shout of joy that burst from the lips of the men who were elated by the triumph alarmed the forces which held the second line, that was parallel with the first that had been taken; and rebels who were subsequently captured stated that their army was aroused and saved by this cheering.[32]

The enemy opened before all the guns had been removed; and the woodwork of those near a salient angle of the fortification, which was the centre of the ceaseless combat during the day, was riddled and rendered useless by the thousands of

---

31 Original text says "Stuart." Confederate division commander Edward Johnson, and one of his brigade commanders, George H. Steuart.

32 The II corps attack had Barlow's division on the left in front, with Birney on his right; Gibbon was behind Barlow and Mott behind Birney. Lt. Col. Burns' (Excelsior Brigade) noted the capture of "150 prisoners, 2 strands of colors, and 2 pieces of artillery, one of which was turned and used against the enemy with great effect. The command succeeded in getting the prisoners, colors, and pieces to the rear, but being entirely unsupported, and the enemy concentrating his whole fire upon it, the works so gallantly won had to be abandoned." [OR XXXVI/1, 505]

bullets that were aimed by both armies. The regiment was temporarily detached to assist the provost-guard at a certain point; and all stragglers, without regard to excuses, and those who bore wounded officers, comrades, and even brothers, from the works, were halted, and obliged to rejoin their commands. The situation of the disabled, who were exposed within range of shells and cannon balls, was made heart-rending in the extreme by the absence of the ambulances and stretcher-bearers who should have carried them from the field to the hospitals, where their sufferings would be alleviated. The prisoners were ordered to convey them to the rear, and gladly hastened to discharge the duty, because their lives were imperilled as long as they remained in the front. The heavy shower which fell at eight, A.M., ruined the cartridges in some muskets, but did not stay the work of carnage, which continued until sunset. At this time the author was detailed by Gen. Mott,[33] the faithful commander of the division, to reconnoitre the position of the foe with his company, and report the strength of the force in front, which was invisible on account of a slight elevation that arose between the first and second lines of breastworks. While I was reading at 8.50, A.M., the inscription upon a large flag, from a point of observation that had been gained with ease, my right thigh was affected by the sensation that follows a sudden blow: the muscles

––––––––––––––

33 Blake recalled the incident in his memoirs: "General Mott knew me and
shouted, 'Take a dozen men, go up that hill and report what is behind it.'
There were no pickets in sight, but, when we started, they jumped from the
ground and underbrush, and, when we went over the top, they were
running down the south side of the hill. Their breastworks protected about
6,000 men; their flags were blowing in the breeze and I was at once
conscious of the fact that my small force was between two great armies.
The Rebel sharpshooters at once demonstrated in a practical way that we
were within range of their rifles. Two of my men were wounded, and I was
wounded in the thigh." [Old Soldier]

of the leg instantly contracted; and I was surprised when I discovered that a bullet had ploughed through the flesh.

My steps to the rear were necessarily slow: the reserves, the headquarters of the corps, the sentinels of the provost-guard, and squads of non-combatants, were passed; and more than two hours elapsed before I arrived at the division hospital, which was two miles from the field. Gen. Hancock was issuing orders to arrest the cowards who were constantly escaping from the front, and exclaimed, "These skulkers wish to enjoy the fruits of victory, but are unwilling to share the dangers to win it." The arrangements for the treatment of the disabled were most excellent: a board of experienced surgeons held a consultation upon every case in which amputation took place; and all that medical skill and attention could effect was readily performed. The Government supplies were abundant; nourishment of every description was bestowed; and faithful nurses often brought the cold water, which was Nature's restoring liniment, and saturated the bandages. A small strip of white cloth was fastened to the button-hole of the coat as soon as the names of the wounded were recorded; and the sufferers of every rank and both armies received the same kind treatment, and reposed upon beds of pine boughs in the capacious hospital tents. More than three-fourths of the number were untroubled by pain; and one man who had lost a leg remarked, "I should think that my foot was on, for I have a queer feeling in the ankle:" another replied to this as he raised the stump that had once been the right arm, "I have the same feeling in my wrist which you have in your ankle." The rebels frankly admitted that their wounds were better dressed than they would have been if they had not been taken prisoners; and many amicable conversations ensued between those who had been rendered helpless while engaged in the deadly combat. Strange as the statement may appear, the rank and file always expressed the same opinions; earnestly wished to see a united country; indulged in contemplating visions of its strength; and portrayed the resistless power with which the ablest officers, North and South, leading their commands of veterans in a common cause against the English in Canada and the British Provinces, and the French in Mexico, would sweep them into the ocean and the gulf.[34] Those who belonged to Steuart's[35] brigade evinced a deep hatred towards him on account of his tyrannical conduct, and hoped that he would be treated in the harshest manner by the Union troops. They said, that, when the batteries were hurling solid shot against their breastworks upon the 11th, he coolly shouted, "They have thrown balls enough: I hope they will send some chains; and then I can fasten them to the legs of my men, so that they cannot run away."[36]

The correspondents of the newspapers eagerly questioned the staff-officers to ascertain the details of the battle which they had not witnessed; and by this means I obtained a knowledge of the origin of many untruthful items, — that Gen. This saved the day at one point, and Gen. That at another time turned defeat into victory. A large number of skulkers concealed themselves in the forests, or bivouacked near the hospitals, and feigned wounds by binding up their heads and arms in blood-stained bandages, or limped, with the assistance of a crutch, in apparent pain; and details of the provost-guard frequently patrolled the ground to seize these base wretches, and escort them to the front.

---

34 The French had installed the Hapsburg Emperor Maximilian in Mexico while the Civil War was raging in the United States.

35 Original text says "Stuart."

36 "Maryland" Steuart was not "treated in the harshest manner" by his Union captors – he was exchanged several weeks later and returned to action with the Army of Northern Virginia through Appomattox, where he was one of the officers that accompanied Lee during the surrender.

The army thieves, who lurked in the rear and waited for the cessation of the conflict before they plundered the slain, grasped with their remorseless hands the valuables, clothing, and rations of the unwary, wounded soldiers, the flattened bullets that had been retained as priceless relics by those from whom they were extracted, and the invaluable swords which officers had borne with honor through scenes of carnage. In the tent to which, with twenty others, I was assigned, a member of the regiment was robbed of every thing, including an old knife and a diary, while he was unconscious on account of a ball which entered his head; and another person was plundered in a similar style before he had recovered from the effect of the ether which had been administered when his arm was amputated.

The heavy musket-firing, which continued throughout the night, ceased on the morning of the 13th, and quietness reigned until half-past eleven, A.M., when the first cannon opened in a slight engagement. Boughs and leaves were placed in the wagons that were proceeding to Acquia Creek for ammunition, forage, and army supplies, to make them suitable for the transportation of the wounded; and in the afternoon the lengthy trains of the ambulances, accompanied by hundreds on foot who were not severely injured in the head and upper extremities, moved upon the plank road to Fredericksburg. The conduct of unscrupulous agents, who acted in behalf of societies that induced the Government to allow them to supply the wants of the disabled at this point, caused a scarcity of food, lack of attendants, and universal suffering, to such an extent that many clamored to be sent to the front; and a shout of joy was heard when the marching orders were received. Nothing could be purchased in the desolate city, which had remained unaltered since the bombardment, and consisted of empty storehouses, deserted buildings, and a small number of scantily clothed and fed citizens. The chaplain who had been dismissed from the service for stealing a horse was very active, and circulated Bibles and tracts for the Christian Commission among the wounded. The ambulances were slowly driven through the fields upon which the army had encamped at Falmouth, and reached Belle-Plain Landing at midnight on the 15th. Four thousand rebel prisoners, awaiting transportation, were confined, near the mouth of the Potomac Creek, in a natural basin that was enclosed by high hills upon which were posted guards, and guns loaded with canister. The wounds of all were examined before they were allowed to enter the transport, to detect the cowardly imposters, a large squad of whom was sent under guard to the detachment of skulkers, which numbered a thousand. A body of twenty shirking officers, some of whom were in irons, who had escaped from the battle of the Wilderness to Washington, marched by the ambulance; and I could not imagine a stronger cause for the emotion of humiliation and shame which was expressed by their dejected and averted faces. The steamer that had been fitted for its use with commendable foresight before the army crossed the Rapidan was amply supplied by the Government with every article that was required for the shattered frames of its passengers.

Chapter XVIII

The Hospital, and the Treatment of Army Diseases

THE STRENGTH of the wounded was completely exhausted in Washington by waiting upon pompous and unfeeling officials, who viewed with contempt the men that performed the fighting in the front while they flourished in luxurious ease and safety in the rear. The surgeons in the office of the medical examiner ordered me to report to the hospital at Annapolis for the purpose of obtaining a leave of absence, which they said they could not grant on account of the large number of cases that demanded their services. A friend who had been crippled in the army remarked in emphatic terms, when I communicated this fact to him, "Don't go: it is all made up between them and the doctors in Annapolis; they lied to me in the same way when I was sent there." I again reported for instructions, to prevent any mistake; and they asserted in the most positive language to my comrade, a wounded officer of the regiment,[1] "You will get your leaves as soon as you arrive at Annapolis." An ex-lieutenant-colonel, who had served upon the staff of a commander in the Army of the Potomac, said to me when I left the hotel in the evening, "I am very sorry that I didn't see you when you came: I should have introduced you to ______, who has the papers put through, and gets twenty or thirty every day. It would cost you a trifle for his trouble." The captain of a New-York regiment procured without delay a leave from Secretary-of-War Stanton, through the influence of an M.C. who was a notorious enemy of the Government; and an officer from the same State entertained the

---

1 Capt. Alexander McTavish, who later returned to the regiment and was killed in action at the battle of Boydton Plank Road, October 27, 1864.

crowd by saying, "This was the style when I applied for my document;" and then limped with groans, and a countenance upon which extreme agony was depicted. "This is the style now that I've got it," he uttered with an animated laugh, as he danced upon the floor and invited everybody to walk to the bar.

The brick buildings that had been used by the students and professors of the Naval Academy had been converted into hospitals, and I was conducted to a pleasant room or "ward," from which the eye could gaze at the beautiful grounds and the Chesapeake Bay. My astonishment may be imagined when a surgeon informed me that "we never grant leaves of absence; and you must remain in this place until your term of service expires, because your wound will not be healed previous to that time." He made another consoling assertion: "It is useless for anybody to seek to escape, as Gen. Hancock always disapproves every application which is not sanctioned by the officials of Annapolis."[2] When this conversation was repeated, I found that most of the wounded patients, comprising fifty or sixty in number, had been deceived by the same false promises from the medical authorities of Washington. In reply to a remark which I made to the nurse, he said that he did not know the proper mode of applying bandages, because it was the first day that he had been detailed for this duty; and the visiting surgeon, a most excellent and faithful public servant, stated that their attendants were ignorant, for they had been compelled to detach those who possessed experience for service in the front. Persons who understand the nature of wounds, and the necessity of the best treatment and watchfulness, can realize the sad condition in

---

2   As commander of the II corps, Hancock would have had final say on leaves of absence granted to the officers of his corps. Hancock had his own dealings with the medical department after his wounding at Gettysburg — a wound which flared up during the Petersburg campaign, necessitating his stepping down temporarily from command of the corps.

which officers were placed who must linger for months among heartless strangers when they most ardently desired to return to their homes. Many indignantly protested that their confinement was a punishment for the faithful performance of their duty upon the battle-field. Wines and different kinds of liquors, which were daily prescribed for three of the most helpless victims in the ward, were received in very small quantities about twice a week; and the nurses were frequently intoxicated, and disturbed the suffering inmates by their boisterous conduct.

"Where were you yesterday?" an officer asked: "there was no one to wait on us for twenty-four hours."

"I was drunk, and shut up in the guard-house," the man truthfully replied.

"Where do you all get so much liquor?"

"It don't cost any thing. It comes from the _______," naming, not the Government, but a vast association which is bountifully sustained by the large heart of the patriotic North.[3]

This hospital, like most of those which are located in the rear, furnished a splendid refuge for the skulkers of every rank. Some had lived within its walls two years in perfect health; and I discovered one worthless fellow who had been absent from the regiment from the first battle of Bull Run. A few officers openly arrayed themselves in the apparel of civilians, without receiving a reprimand, — although it was contrary to explicit orders from the War Department, — and enjoyed the privileges of the town, which were more agreeable to them than the display of courage which defies the bullet and rifled cannon. The most pitiable class of patients consisted of prisoners that had been recently released from the horrors of Libby Prison [in Richmond, Virginia]; and some, whose minds had been temporarily weakened when their

---

3 The two largest relief agencies were the United States Sanitary
   Commission and the United States Christian Commission.

frames became emaciated, talked and played together with the simplicity of youth. The untiring exertions of Hon. Oakes Ames,[4] the representative in Congress from the district, overcame every obstacle; and I succeeded in procuring the much-desired leave of absence from the Secretary of War upon the 27th [of May], after a vexatious controversy with the chief surgeon;[5] and no Gen. Halleck or medical director could withhold it.[6] Another surgeon, to whom the "almighty dollar" per day was paid for board, stated that an ambulance would be furnished to convey me and my comrades to the railroad depot; but none was supplied, although there were ten or twelve which were parked in the stables. The evening train was thus lost; a delay of twelve hours was caused by this broken promise; and upon the morning of the 28th I gladly quitted the scenes in which my brief sojourn had been so gloomy and unpleasant.

The result of the experience and personal observations of the author, and the unbroken testimony of those who had been disabled in the service, support the disgraceful facts which are recorded in this paragraph. The convalescent and parole camps and the permanent hospitals that were established in the vicinity of the cities of the loyal States, with a few honorable exceptions I

---

4   Ames was Representative to the U. S. Congress from Massachusetts, 1863-73. After the war, he was a share-holder in Crédit Mobilier corporation, which obtained contracts for supplying the construction of the Union Pacific Railroad. By milking the railroad with exorbitant prices, Crédit Mobilier's dividends climbed more than 600 percent. Bribes paid to congressmen resulted in laws giving tremendous land grants and liberal loans to the railroad. The scandal broke in 1872 and Ames was censured by the House.

5   The acting Surgeon-General had returned Blake's application for leave, noting that "The statements of Lieut. Blake and McTavish are not substantiated, and leave of absence is not recommended." Apparently Ames' intervention prevailed. [NARA, Henry Blake File]

6   General-in-Chief Henry W. Halleck was the universal scapegoat for almost any administrative foul-up.

trust, have been the centres around which deeds of iniquity revolved; and a majority of the surgeons connected with them have been base vampires, that exhausted the vitality of the Union armies in the darkest hours of this struggle.[7] Their lists of able-bodied soldiers whom they falsely reported upon the sick-rolls grew larger and larger; while the regiments in the front, in the same ratio, became smaller and smaller. More than two-thirds of the hirelings styled nurses and servants, that were employed in these institutions, were cowards and skulkers of the vilest order, who ran away from their comrades in the day of battle, and gained the favor of their medical officers by the most abject servility. When the infamous system began upon the Peninsula, the company commanders refused to forward papers which the surgeons demanded to enable them to defraud the Government by receiving the pay that is due to soldiers; and Secretary Stanton vainly menaced them, "by order of the Secretary of War," with threats of dismissal from the service for neglect of duty. That bond of wickedness, the "partial descriptive list" [of sickness or disability], was then framed and issued by the departments, by means of which millions of the national treasury have been regularly paid to thousands of deserters, who were sheltered from arrest by the medical directors, who were sustained by the powerful authority of the Secretary of War. Hundreds of cunning malingerers lurked in these secure retreats;[8] and noble men who

---

7   Blake's caustic remarks about the the convalescent and parole camps and
    the permanent hospitals, and his comment that the majority of surgeons
    connected with them were "base vampires" are probably fairly
    representative of the feelings of veteran combat officers of understrength
    regiments.

8   George W. Adams, who wrote a medical history of the Civil War, saw the
    problem in a different light: "All that can be said in favor of the use of
    convalescents as nurses is that it saved some men from premature return to
    the lines and gave them something to do. … To relieve their boredom a few
    pulled wires to rejoin their regiments, only to suffer relapses or crawl

had been wounded were not allowed to return to the front when they had recovered, although they sometimes applied eight or ten times for permission, and at last rejoined their commands by eluding the vigilance of the guards. Soldiers of the last honorable description could be found in every company of the regiment; and the officers always commended their conduct, and tore into shreds, or cast into the fire, notices stating that they were deserters from the so-called hospitals. Delicacies and all that satisfied the appetites of the body, which had been abundantly supplied by the philanthropy of the American people, were usually stolen by these miscreants; although I admit that they never withheld or retained for their private use the tracts and Bibles which had been presented as food for the soul. The author most cheerfully excepts from this severe criticism the female nurses, who performed their tasks, in the few places from which the malignity of unscrupulous surgeons could not exclude them, with a tenderness and honesty that secured for them the esteem of every person who was the fortunate subject of their attentions. The substitution of these high-minded attendants of the gentler sex for the diabolical wretches who should be transported to the front would re-enforce the army, and encourage those that meet the enemy upon the field of conflict.

---

through months of unhappy semi-invalidism.

"Colonels often complained that hospital held patients too long, but the evidence points to the opposite. The overcrowding of hospitals early in the war led to overhasty discharges. Adams points to the varying results of the special commissions which were sent to the hospitals to determine which patients might be returned to duty: Commission 'A,' headed by a medical inspector, left 23 percent in the hospital; Commission, 'B,' headed by a line officer, only 3 per cent. But surgeons-in-charge soon learned how to fight back. They simply 'forgot' to send back a patient whose health might be endangered by removal. And attendants whom they valued were simply hidden when the examiners appeared." [Adams, 186]

The policy thus accurately described was so generally carried into execution, that it was a maxim among officers and men, that no soldier who entered a hospital in the rear ever shouldered again his musket with his comrades in the ranks. Discharges from the service were often issued to those that were able-bodied; and, upon one occasion, the passenger-car in which I sat was filled with these knaves, who uncorked their bottles of stolen liquors as soon as the train started from Washington, and publicly boasted that they had never seen a fight, but "beat the Government" one or two years, and bought their final papers for certain amounts of money. A brief career of dissipation exhausted their funds, and many of them rejoined the army within a month after they had left it. A member of the company skulked from Williamsburg, and joined a hospital; which alarmed his wife, who supposed he was seriously ill, and wrote letters of anxious inquiry to the regiment: but he informed her that he was feigning sickness to escape the dangers of battle, and pacified her by the pleasant suggestion that he could not be killed. The surgeons aided the shirk in every way, and applied about once in two months for his descriptive list during the ensuing two years, although they were continually notified regarding his real character; and finally granted him an honorable discharge. The topic of malingering is endless; and many incidents might be narrated to illustrate the various "dodges," of means of "playing it." The tricks of European soldiers were revived by foreigners from the Continent; and the regimental surgeon detected the copper which produced a frightful sore upon the knee of one of the company by noticing the impression of the American cent. Others pretended to be afflicted with every disease that troubles the human system; but the frequent prescription of the most nauseous doses sometimes checked their complaints. A private was excused from performing any duty for three months on account of rheumatism; and always walked with great difficulty by the aid of a cane, and daily expected to receive a discharge for disability. Unfortunately for his well-laid scheme, he foolishly became intoxicated, and appeared upon the parade-ground without any artificial means of support, and ejaculated to the astonished crowd, "I ain't lame; I'm playing it; I don't need any stick." He was detailed for guard upon the following morning; subsequently deserted from the regiment at Fair Oaks; and, like thousands of similar rascals, obtained an honorable discharge from the hospitals. "That's the way I got it," he remarked as he chuckled, and slapped his hands upon his pocket in a significant manner.

Fingers were sometimes shot off, and other wounds were self-inflicted, to attain this object; and in two instances a serious miscalculation resulted in the loss of a foot and an arm. A substitute of the adjoining regiment persisted in carrying a cane upon the marches, reviews, and inspections which he was obliged to attend, and declared that he could not walk without using it; but at Locust Grove the bullets frightened him, and he ran from the woods with a speed that was seldom excelled. The uniform of the invalid corps,[9] which was wisely designed for the

---

9   The Invalid Corps (IC), formed in April 1863, had a sky blue uniform. It was composed of two battalions, one of men fit for garrison duty and the second of amputees who might be able to serve as watch-men or hospital attendants. The corps grew to 28,000 men during its eleven-month existence, three-fourths of whom in the first battalion of the physically able.

   Recovering soldiers were not attracted to the IC, because those who felt strong enough wanted to return to their regiments; if they felt weak they wanted to be sent home. Many disliked the IC because its members were neither soldiers nor civilians, and felt they would be branded as cowards.

   The IC reorganized as the Veterans Reserve Corps (VRC) in March 1864, which attracted less criticism. It had a battalion of expired three-year men who did not want to re-enlist for combat duty. They served as railway and bridge guards. The second battalion "gave the Medical Department

most exalted purposes by its authors, is disgraced by a majority of the skulkers that wear it; and the number of the disabled in its ranks who could — *Count the dates of battles by his scars* — is very limited. A multitude of worthless officers and men, enfeebled by the "cannon-fever," rushed into its regiments, which in physical vigor were often superior to those that labored in the trenches or fought the enemies of the country. The surgeons once more decimated the national forces by ordering the names of those who were asking for permission to return to their old commands to be enrolled as members of this corps. Its reputation was so seriously impaired by these practices, that soldiers of honor and principle, who had been crippled in the services, refused to enlist in it, because they knew that their military fame would be tarnished by the inglorious action of their associates.

The ambulance system, and the means provided by the Government for the speedy removal of the wounded, were most excellent; but the evasion and the negligence of the agents to whom they were usually intrusted, rendered them, at times, valueless. Officers and men of doubtful courage used every exertion to be detailed for this service, because they considered it a safe position; and, with rare exceptions, remained in the rear while the wounded were stretched upon the field and praying for assistance. The regimental commander [Blaisdell], upon one occasion, issued the following order to the company officers when he received a notice to detail ten soldiers for duty in the ambulance corps: —

---

what it had long wanted — organized military companies, controlled by line officers to guard the hospitals and care for the clothing and effects of the patients." [Doctors in Blue, 187-188]

"Take the most worthless cowards and stragglers that you have got: I won't insult my good and brave men by sending them to such a lot of scalawags."

There was scarcely a day upon which the wagons were not used for some foreign purpose; and, upon the march to Gettysburg, a general in the corps[10] appropriated three of them which conveyed sumptuous stores of luxuries and liquors, and retained them when the Dutch farmers were filching money from the helpless whom they transported a short distance. When the drivers bivouacked upon the road to Falmouth, they compelled the sick to leave the ambulances in the midst of a pitiless storm; and commanded them to go to their regiments, which were five miles from that point, because they wished to sleep in them during the night. One victim of this heartless cruelty, unable to walk, and overcome by the state of despair that enveloped him, finished his life with the musket that he had so often aimed at the ranks of the enemy. A formal complaint, which recited these facts, was forwarded to headquarters by the regimental surgeon, but the only notice that was taken of it was the extraordinary answer, that such conduct was customary, and in accordance with standing orders. The appointment of brave officers and men for this department, throughout the army, would save many valuable lives, and silence the objections that have been publicly urged against the corps.

The hospitals in the front, without many of the external conveniences, rank high above those in the rear, because they were usually managed for the best interests of the service; although some of the abuses that always follow the employment of non-combatants in the field were occasionally developed. The

---

10 Probably referring again to Joseph B. Carr, or possibly Andrew A. Humphreys, who Blake held in equally low esteem in terms of serving his personal needs over those of his men.

presence of comrades who made frequent visits to assist the suffering, the superior care of friendly attendants who did not wish to pilfer, and the bands which were detailed to play [music] upon certain days, had a beneficial effect. Gen. Hooker entered every ward at Falmouth after the battle of Chancellorsville; remedied all the defects that were visible; conversed with the wounded; and wrote a pass for the mother of a soldier who had hastened to Washington, but, unsupported by influence, was unable to reach the couch of her dying son. Every means were used that conduced to preserve a cheerful and contented disposition, which decided the issue in doubtful cases in favor of life; for what is termed homesickness has caused the death of many soldiers. Mangled men played games of cards or checkers; those without a leg proposed to dance or race; others, without an arm, challenged a comrade equally helpless to box or wrestle; incidents of the battle were described while the tobacco in the pipe was unconsumed; grotesque and useful articles were ingeniously made; and the soldiers sometimes carved into ornaments the bones which had once formed a part of their bodies.

The effect of wounds upon different constitutions, in the excitement of the conflict, was very striking; and those with the slightest injuries frequently exhibited the utmost distress. An officer who bivouacked with the regiment at Gettysburg, and was certain that he had not been struck, and walked unassisted to the hospital, because his "side felt so sore," was not more amazed when the surgeon informed him that a bullet had barely missed his lung, than a colonel whose uniform had been perforated by balls, and who was borne upon a stretcher from the field, after exhorting his men to boldly face the foe, when he was told that there was not a scratch upon his person. The most heart-rending scenes that I ever witnessed in my eventful experience were those in which the helpless vainly and piteously implored their comrades to shoot them, and end their excruciating agonies. Careful habits promoted the health of the soldiers; and the rules of army mortality were reversed in many regiments (including the 11th) that performed severe fighting; and four or five men died by the casualties of battle to one that perished by disease. The roar of artillery, and the sudden shock of conflict, occasionally produced deafness or speechlessness; and the concussion of a shell sometimes killed a man, or fractured a limb, without inflicting a wound. The bullets rarely traveled in the same grooves through the air; but an officer was struck in the same place in his jaw at Fair Oaks and Fredericksburg, and one soldier received four wounds in the left arm in different battles. Anxiety and responsibility, which were suspended over the head like the sword of the ancient emperor, produced premature old age: many generals soon lost their youthful looks; and it was not uncommon to see gray-haired veterans who had not lived twenty-five years. The proximity of battle always affected the health of certain soldiers; and there were persons of high rank that were sick upon such occasions, and never exposed themselves within the range of rebel cannon. There is a class of generals that are never attacked by disease whenever they are engaged upon court-martial, provost-guard, and similar duty at Washington, or the safe places of the North; but an order to report to the front is always succeeded by prostration and weakness.

Statements that "only forty men are left in a regiment that once numbered a thousand," and the exciting and exaggerated accounts about the havoc of battle, have led many to largely over-estimate the number of those that die in the service. The long list of the deserters, the discharged, and the non-combatants, that sometimes include one-fourth of the aggregate strength, explains the nature of this apparent waste; and there were few of the three-years' regiments which lost more than ten officers, or two hundred and fifty men, by death from all causes.

Chapter XIX

Army Morality[1] and Discipline

MANY FACTS which have been narrated in the foregoing chapters might be classed under this head; but a separate discussion of the subject has been deemed of importance. The religious belief of the army was simple, and consisted of two articles of faith: first, that "a man will die when his time comes;" and secondly, that "a soldier who is slain in the service of his country is sure to enter the gates of heaven." The arguments of books and the sermons of divines could not undermine these ideas, in the sincere profession of which thousands fought and died upon the battle-field. The chaplains of the army, those that should be the types of its purity, were commissioned without regard to their moral qualifications; and, as a class, exerted a debasing influence upon the soldiers: so that it was generally impossible to perceive any distinction between the man of God and the man of sin.[2] The officers of some regiments, from which they [the chaplains] had been dismissed for military offenses, voted that it was inexpedient to procure another spiritual adviser, because they considered that his example would be as pernicious as that of his predecessor and brethren in the service. Tracts upon the wickedness of dancing, attending theatres, sleeping in church, extravagance in dress, and similar matters, were extensively circulated among the troops; and it was evident that the dealers

---

1 For an extensive discussion of gambling, drinking, and the use of profanity during the Civil War see Wiley, 247-255.

2 Blake did not have a favorable opinion of army chaplains generally, however, Warren Cudworth of the 1st Massachusetts was an exception: Cudworth had been a minister at Blake's Unitarian church in Dorchester and was a fellow abolitionist.

had shrewdly cleared their shelves of the unsalable rubbish which had been accumulating for years. The letters that were frequently published in the religious papers portrayed the marvellous results which ensued when they were distributed, perverted the facts, and deceived those well-meaning philanthropists that contributed them. Less than five soldiers in the regiment perused them with a conscientious interest; and this small number of readers was not largely exceeded in other commands that passed within the range of the author's notice. I have often witnessed the following incidents when a certain chaplain appeared in the company streets to give away the tracts which had been consigned to him. A squad of men, jostling each other, and using many oaths, surrounded him, and shouted, "I don't want them little things: give me some of those big papers with the flag on them. I am going on picket, and want some to put my rations in." "Those tracts are just what I want to light my pipe with." "Give me some too: they are first-rate to kindle fires." The chaplain gratified the request of each person; had a jocose answer for all; and said, with many repetitions, "Use them for any purpose you see fit. It is my business to get rid of them; and it is nothing to me what you do with them." If they asked for some flashy novel, he replied, "I have not got it now, but will let you have it if it is sent to me."

The week that succeeded pay-day was noticeable, because groups of men assembled in the vicinity of the regiment, and gambled hour after hour with cards, dice, and props, and rubber blankets, upon which certain squares and digits had been rudely sketched. The number of players gradually diminished; and the few winners, who had acquired the stakes which had been lost by the majority, contended for heavy amounts, and one hundred dollars were frequently placed upon the board. Some gained by this means large sums; and a soldier at Culpeper Court House accumulated one thousand dollars in the course of twenty-four hours. It was usually prohibited by officers, and guards sometimes patrolled the camps to enforce the order; but others did not interfere: and the brigade commander remarked to the division general when they inspected the grounds which the gamblers occupied, "God Almighty never made a better place than these woods for men to gamble in." The passion displayed itself upon every favorable opportunity; and the implements of the various games were actively used upon the march, the picket-reserve, and the battle-field. A chaplain, who acted the part of a spectator, innocently inquired, "How does this man take that man's money?" The problem was quickly solved; and whenever he was invited to join, or "take a hand," he declined, not by uttering stern reproof, but some facetious remarks, "I have two hands now, and don't want another;" or, "I am afraid that I should lose my money if I played with such skilful experts." The "lucky ones" generally rolled their greenbacks together like a twist of natural leaf-tobacco, and forwarded them home; although some honorable chaplains refused to assist any gambler in saving his gains, and delivered strong addresses against the alluring vice. The conduct which has been censured relates wholly to a class that comprised about one-tenth of the enlisted men in a regiment; and, with the exception of a solitary occurrence in another brigade, I never observed an indiscriminate medley of officers and soldiers in the crowd of players.

Profanity, which was unchecked by the presence of the restraining influence of home and a civilized community, was a habit that existed in every rank and grade; and the old saying, "to swear like a trooper," was hourly confirmed. The language generally used in the summons to surrender, by the members of both armies; the actual terms employed by officers of high rank in giving orders; the exclamations of the wounded, and the last words of dying heroes, — seldom appear in the volumes of the historian.[3] It was an interesting study to observe that foreign recruits, who were unacquainted with the English tongue when they entered the regiment, quickly acquired a proficiency in the use of the strongest oaths before they had mastered the rudiments of their adopted language. This discreditable feature of the army always struck the attention of visitors by its publicity; and many thoughtlessly uttered another ancient maxim, "The worse the man, the better the soldier." A brief military experience will satisfy all concerning the falsity of this remark. The worthless bully and idler in the abodes of peace invariably form the most useless parts of the engine of war; while the quiet and industrious civilians constitute the motive power and essential portions. A lofty principle of action, not physical brutality, is the basis of that courage and heroism which are absolutely necessary to insure the success of the subtlest planned campaign. I have seen the pugilist who feared no antagonist in the wrangles and scuffles of a mob tremble with fright, and flee, upon the field of carnage while a youth animated by pure patriotism, who timidly shrank from the quarrels of the street, was the foremost in the charge of victory, and the most steadfast in the severe trial of defeat.

"I was brought up on a bottle, and never saw any harm in it," was the observation of a drunken chaplain, that furnishes a text for this paragraph.[4] The enforcement of stringent orders

---

3 For officer swearing as an advanced art form see the discussion on Gen. Humphreys on pp.44-53, 325, 329 in Red Tape.

4 In addition to other character flaws pointed out earlier by Blake, it seems Chaplain Watson had a weakness for alcohol, as noted in the *Boston Saturday Evening Express*, February 22, 1862: "The Rev. Elisha F. Watson, who for some years was a Methodist minister, ... was appointed last spring to the chaplaincy of the 11th Massachusetts regiment, and has accompanied it from that time to the present, and was at the Bull Run fight. He is now under arrest for conduct unbecoming a minister and a Christian, the principle charge against him being for getting drunk."

prevented the sale of liquor to the rank and file, so that a compulsory sobriety existed; and the number of intemperate soldiers who bore muskets with their commands was extremely small. Some, who foolishly supposed that this stimulant was indispensable, admitted their mistake when they were compelled to perform fatiguing labor without it; and the army in this way aided a great reform. Venal officers, sutlers, and commissaries, made enormous profits by covertly selling the forbidden beverage; and privates sometimes willingly paid a hard-earned wages of a month to procure a canteen of whiskey which cost the vendor twenty cents. The course of the Government and charitable associations in issuing it as a ration, in certain circumstances, to the enlisted men, produced evils that are boundless in their extent. The well-known demoralization and inefficiency in battles of many German regiments, especially those that "trinks mit everybody and runs mit Howard," were caused not by the Teutonic but the lager-beer blood that pervaded their systems. I express in print what has been said many times in conversation by officers, that the total prohibition of the use of intoxicating liquor for any purpose by any class of persons in the army would have preserved the lives of thousands, and shortened the duration of this war by at least one year. The careful reader has noticed that the shameful drunkenness of a corps commander[5] became the stumbling-block in the path to victory, when Gen. Meade was foiled in the movement which terminated at Mine Run.

A broad distinction was established; and it is a sad fact, that, with rare exceptions, the commissioned officers were not only unhindered, but even aided, to obtain the ardent spirits which they desired; and every brigade commissary was supplied by the

---

5 Major General William French, whom the troops called "Old Blinky."

Government with barrels of whiskey for this purpose. It was a sharp device, at one time, for privates to put on shoulder-straps, and purchase liquor, which was delivered to them upon the supposition that they were officers. The tents of may generals and subalterns presented the appearance of the glittering and highly-colored glasses and bottles of bar-rooms; and the presence of one of them, upon a visit or special duty, was the signal for the production of "something to take." From the first battle of Bull Run, at which Miles[6] was allowed by Gen. McDowell to disgrace himself and the nation with impunity, to Spottsylvania Court House, where I counted, a few minutes before I was wounded, twenty-six general and staff officers that rode upon their horses with great difficulty on account of intoxication, my minutes contain notices of drunkenness upon every scene of conflict in which I was engaged. To avert suspicion, I desire to state, that none of the officers with the regiment were rendered inefficient by this cause, in such important crises.[7] The practice existed to such an alarming extent, that, when certain persons fell in the time of action, it was a subject of doubt, with the distant spectators, whether the cause was a bullet or the contents of a flask. "I need three canteens a day now," a staff-officer remarked during the last campaign. The rebel prisoners with whom I conversed admitted that the same state of facts prevailed in their army, and mentioned the names of some of their commanders who were notorious drunkards. The rights of subordinates were disregarded to pamper this debasing appetite; and a general seized the mail-wagon of the division upon a long march, and retained it more than two weeks (during which time the soldiers

---

6 Col. Dixon S. Miles was relieved by General McDowell at First Bull Run for apparent drunkenness.

7 Col. Blaisdell was well-known in the army as a drinker, but also had a reputation as a fighting colonel.

received no letters), in order that his stock of liquors might be transported, — a fair quantity of which he openly imbibed upon the battle-field. I was stationed upon picket when the ambulances returned from Chancellorsville, which they had visited by means of a flag of truce; and men of veracity, who belonged to the regiment, assured me that many of the surgeons were "tight;" and they preferred to keep on their bandages of shelter-tent, which were black with clotted blood, and allow their wounds to remain undressed, because they did not wish to submit to the care of such miscreants.

Deserters formed the largest class of criminals in the army; and nearly every regiment that has been posted in the front lost more men from this cause than the aggregate of all the others. The wholesale villainy of the bounty-jumpers has doubled the names that are recorded upon the rolls of dishonor.[8] The friends of the rebel conspiracy in Alexandria, Washington, and other cities, afforded all the assistance in their power to men that wished to escape from the service; and furnished money, suits of clothing, and forged passes, discharges, furloughs, and similar documents, whenever the necessities of the case required them.

---

8 Conscription and the allowance of substitutes and bounties exacerbated desertion rates. The apprehension of deserters was accelerated as the war went on. The number of deserters arrested and returned after April 1, 1863 until the close of the war was about 76,000. The Veterans Reserve Corps proved useful in apprehending and guarding deserters in the 1864-1865 period. Execution of deserters was also greatly increased:

"A gallows and shooting-ground were provided in each corps and scarcely a Friday passed during the winter of 1863-1864 that some wretched deserter did not suffer the death penalty in the Army of the Potomac. During the latter part of 1864 the death penalty was so unsparingly used that executions were of almost daily occurrence in most of the armies. [Lonn, 181-182] Among Massachusetts three-year infantry regiments the 11th was second only to the 28th in desertion percentage at 17.8% (the 28th had 18.2%). [HDS]

The subject of crime naturally leads to a consideration of military discipline, and the punishment inflicted for the commission of offenses of different degrees. The constant interference by generals of high rank, and inter-meddling officials in Washington, have often seriously impaired the efficiency of the troops, by preventing the impartial administration of justice, and the expiation of penalties which should follow a just sentence by court-martial. The first abuse [generals] could be easily remedied by a revision of the articles of the military code which relate to the measures that precede the trial of the prisoner; but the second [politicians] cannot be corrected. Charges of a serious character must be approved by a general officer, who may deliberately suppress them without any regard to the just interests of the service, if the culprit is one of his friends, or can repay him for his sinful kindness. Specifications that were filed against the brother of a division commander[9] for habitual drunkenness never emerged from their hiding-place when they reached his headquarters. A medical director issued an order that an officer should be arraigned for feigning partial blindness to procure a discharge;[10] and although he had been fined by a court-martial for cowardice, and a complaint for drunkenness had also been preferred, they were promptly cast aside, because he perjured himself at the trial of some parties who had incurred the displeasure of the general. Charges of cowardice which were submitted by one division commander against another were repeatedly suppressed because the guilty person was a personal admirer and flatterer of the head of the army; while the subalterns, who said that they could prove them in any court, were summarily dismissed from the service

---

9   James H. Carr of the 11th New Jersey, brother of Joseph B. Carr.

10 Lt. Alfred DuPaget, 11th New Jersey. [Mutiny, 70]

for the use of disrespectful language.[11] Generals who were inebriates, poltroons, or traitors, were seldom if ever punished, but promoted, to demoralize, by their ignoble conduct, the unfortunate brigade, division, and corps commands which were obliged to serve under them. Recommendations from officers of equal of higher rank triumphed over the just objections of subalterns and enlisted men, and gained a confirmation by the Senate.[12]

The enforcement of the death-penalty against every deserter would have retained thousands in the army, and had an excellent effect upon its discipline; while the clemency that was shown towards them actually encouraged soldiers to commit this infamous crime.[13] Many of the vilest substitutes, who enlisted with the intention to escape from the lines, were pardoned by the authority of Washington dignitaries, although their officers reported that there were no mitigating circumstances in their cases. One of them openly declared to his comrades that he should desert, and no power could injure him, because Senator _______ of Wisconsin and his father were intimate friends. He ran away at the end of a fortnight; was apprehended, and sentenced to be shot by the unanimous vote of the members of the court-

11 Referring to Gen. Prince's charges against Gen. Carr for his actions at Mine Run. One of the subalterns was Capt. James R. Bigelow, 11th Massachusetts, who was court-martialed and dismissed. [Mutiny, 53-66] Blake commented after the war: "Common sense should be appealed to in fixing punishment by a Court Martial; where there is discretion, and an officer who fights, fights, fights should be kept in the service unless his offense is rank and smells to heaven." [Memoirs]

12 Both Carr and Humphreys were ultimately confirmed by the Senate although each appointment was challenged by subordinates.

13 Blake's view on deserters was shared by many combat soldiers, including most of the West Point generals. Early in the war executions for desertion required Presidential review and approval. This was changed by Congress in the 1862-1863 session so that approval of the army commander was sufficient.

martial, who noticed the utter depravity that was visible in his face and demeanor when he boldly said, "I suppose I deserted to swindle the Government." The commander of the company, in answer to inquiries from the War Department, replied that he was the worst soldier that he had ever seen; but the sentence was remitted in compliance with the request of Senator _____ of Wisconsin.[14] A thousand facts like these which have been recited affected the authority of officers by rendering the enlisted men defiant, and at times nearly destroyed the discipline, without which an army becomes a mere assembly of citizens equipped for military duties. The state of perfection which existed in many of the [artillery] batteries that belonged to the regular service exhibited the beneficial results of an implicit obedience to orders. An inflexible command, that was never modified by superiors, moulded the best artillerists in the world from a nucleus of fifteen or twenty members; while the main portion was composed of the most useless and refractory soldiers that could be found in the guard-houses and prisons of the regiments from which they were detailed.

---

14 The Senators from Wisconsin during the war were James Doolittle and Timothy Howe. Blake is apparently referring to a specific case involving one of them, but we have not identified which one.

Appendix

Henry N. Blake

Henry Nichols Blake was born June 5, 1838 in Dorchester, the son of James Howe Blake and Mary Beal Nichols. He was a raised in the Unitarian faith, and attended the public schools of Dorchester, which provided ample preparation for him to attend Harvard Law School, from which he graduated in 1858. He was admitted to the bar on June 9, 1859. Blake described the early years of his law practice in his 1916 memoirs:

I lived at the home of my parents, walking daily, when the weather did not interpose a veto, from Mill St. to the office, in round figures, eight miles. I formed a mental resolution to pass every foot traveler on the road, and if I may be pardoned for boasting, was rarely beaten. Sidewalks were the exception to the rule and the streets were full of ruts and rough.

I did not receive any compensation in money for my services in the office, but was rewarded by practical advice and lessons, courteous treatment by gentlemen, and some good dinners. I had ample time to observe lawyers in and out of court and attended important trials when eminent counsels like Rufus Choate, Caleb Cushing, Benjamin R. Curtis, Richard H. Dana, and Sidney Bartlett were engaged. I listened to a host of lectures, clergymen, and politicians outside of judicial tribunals before I attained my majority: Henry Wilson, Charles Francis Adams, Thomas Hart Benton, George Sewall Boutwell, Anson Burlingame, Nathaniel Prentice Banks, Robert Charles Winthrop, Wendell Phillips, William Lloyd Garrison, George William Curtis, Horace Greeley, Oliver Wendell Holmes, Theodore Parker, Henry Ward Beecher, John Pierpont, Thomas Starr King, and George Putman.

I heard in Faneuil Hall, October 17, 1858, Jefferson Davis, who had a lordly air and was a magnetic speaker, but declared his opinions clearly and concisely and arranged his sentences with logical effect. He steered away from secession and fire-eating topics, and if his career had terminated before the rebellion, his name would not be remembered here.

About this time, I saw an anti-slavery meeting in Music Hall Boston dispersed by a mob of rounders and plug-uglies from the abodes of vice. This is a forcible illustration of the difference between toleration in the north and intolerance of the part of northern men with southern principles.

Blake was an avid supporter of Abraham Lincoln in the election of 1860, marching in the ranks as a member of the "Wide-Awakes" youth organization which formed within the Republican party.

I wore the uniform cape and cap and carried a regulation torch. ... Some nights the march covered eight miles. I did not miss one meeting or parade and the time flew amidst scenes of joviality without an accident.

His support of Lincoln never faltered, as he noted in the *Harvard Alumni Bulletin* in 1927:

I have known voters to express regrets that they had helped to elect candidates who, they later felt, had betrayed them. But from 1860 to the present hour I have rejoiced at every election...that I was a firm upholder of Abraham Lincoln in 1860 and 1864.

Blake abstained from alcohol and tobacco his entire life. Looking back at age 89, he reflected:

Cicero, in his essay on "Old Age," says that a man who desires to attain this great blessing must lay the foundation for good health in his youth. During this important period, I did not sign any document or swear to any pledge, but I formed the mental resolution to follow the example of my good father, who did not drink intoxicating liquors, or use tobacco in any form, and I have stood on this platform and never questioned its wisdom.

Having put his life "on hold" for the three years of his military obligation, and fully recovered from his wounding at Spotsylvania, Henry Blake embraced the credo of "manifest destiny" espoused by Horace Greeley: "Go West, young man, go west and grow with the country." Blake did just that in 1866. After a short-lived attempt at panning for gold in the Montana Territory, he found employment as editor of the Republican newspaper in the territorial capital, Virginia City, and was frequently at odds with the Democratic governor (and former Irish brigade commander) Thomas Francis Meagher.

Blake returned to the law in 1869 as an appointed U.S. Attorney and as an elected member of the territorial legislature.

He made a brief return to Boston in 1870 to marry Clara Jane Clark after a lengthy, long-distance, courtship. In 1875 he was appointed to the territorial Supreme Court, and later became Montana's first Chief Justice after statehood was achieved. He returned to private law practice in 1892, which continued until his retirement in 1910, when he and his wife returned to Boston.[1]

Blake's retirement endured for over two decades, during which he wrote his memoirs and kept active as a public speaker. He died November 29, 1933.

Photo credit: Massachusetts Commandery, Military Order of the Loyal Legion, U.S. Army Military History Institute.

---

1 For a more detailed discussion of Blake's past-war career see *Montana: The Magazine of Western History* (October 1964).

William E. Blaisdell

William E. Blaisdell was born in Alexandria, New Hampshire, November 15, 1815, son of Moses Blaisdell and Nancy McMullen, and an 8th-generation descendant of Ralph Blaisdell, who first arrived at Pemaquid Point, Maine in 1635.

He enlisted as a private in the 4th U.S. Infantry, January 19, 1841. His enlistment paper stated his occupation as "blacksmith," that he had gray eyes, black hair, dark complexion, and stood 5 feet, 6 inches tall.[2] He was promoted to corporal in July 1841 and sergeant in February 1843. During the Mexican War his regiment was in every one of General Winfield Scott's battles. He was seriously wounded during the capture of Mexico City. After the war he transferred to the recruiting service in July 1848, and requested an honorable discharge that was granted in April 1853. He returned to civilian life as an inspector in the Boston Customs House.[3] It was around 1848 that he married Mary Ann Dempsey of Albany, New York. She bore him seven children, most of whom died young, but their youngest daughter lived until 1934. His wife collected a widow's pension until her death in 1908.

At the commencement of the Civil War Blaisdell was offered the rank of captain in the regular army but instead chose to serve in the volunteers, in which he was appointed lieutenant-colonel of the 11th Massachusetts. Blaisdell's background in the U.S.

---

2   Photographs of his 1841 enlistment and "Soldier's Book" can be found on Blaisdell's page at familysearch.org (as of December 2025).

3   Service details are from his widow's pension renewal of 1890.

army must have proved useful during the initial training of the 11th, and his influence may well have been the main reason the regiment performed as well (and as long) as it did at Bull Run.

The first colonel of 11th Massachusetts, George Clark, Jr., had only commanded state militia prior to the war, and while he apparently had the organizational skills and political contacts necessary for raising a new regiment, he fell apart when it came to battlefield leadership. Clark had a questionable performance at Bull Run, and resigned in October 1861. The *Dorchester Beacon* of July 22, 1899 noted that:

> *Col. Clark was never with the regiment after Bull Run. He went immediately to Washington, resigned, and went home. It was not a lack of courage that caused him to do so, although many of the men believed so. He had a very nervous and excitable disposition and worried continually over his affairs instead of taking things as they came. He almost worried himself into brain fever, and was physically unfit to retain his position.*

Blaisdell was promoted to the command he would hold for the next thirty-two months, with occasional short stints as brigade commander.

Col. Blaisdell was well-known in the Army of the Potomac for his drinking and profanity, although he was by no means the only officer with such a reputation. However, he apparently was able to confine his drinking to non-critical situations, as Henry Blake wrote in his memoirs:

> *A military demonstration was made by the Second and Third Corps in February, 1864, and the Colonel of the Regiment was assigned to the command of the Brigade and I acted as his Assistant Adjutant General. The order designated the route and places to be occupied and*

plats of the country were prepared by the engineers for the guidance of the commanders in the march. A driving snow storms chilled the troops, and the Colonel was obfuscated by an overdose of fire water in his efforts to keep warm, but he could sit his horse moving at a slow walk.

A halt was ordered twelve miles from winter quarters near a small church near a fork of the roads, and the Brigade Commander looked silently several minutes at his plat, and then swore about the ignorant fool who didn't know what he was doing when marking the route. I noticed at a glance with my eagle eye that the paper was held upside down, and when it was handed to me, corrected the error by holding it right side up, and showing the highway to a ford on the Rapidan. We bivouacked while the snow was falling fast. ... Resting behind earthworks on the south bank of the river, the serenity on the night was not broken by the rude tocsin of war.

When I awoke in the morning, the Colonel was sitting by the camp fire and studying the plat which persisted in staying wrong side up. His mental condition was unchanged, and he asked, "Blake, how in hell did you hold this thing last night?" I replied promptly by rotating the papers with my fingers. "Blake," he exclaimed, "I always want you on my staff. You are always sober when the rest of us are drunk."

This is an appropriate place to refer to another phase of Colonel Blaisdell and mention his soldierly qualities, his conspicuous bravery, and practical essential knowledge of actual warfare. The test of a soldier is not merely popularity, but the ability to keep cool and direct under fire, and face without flinching the foe and inspire his followers by his example. It is a serious mistake to court-martial and dismiss a veteran of this caliber for a violation of a rule of discipline when a lighter punishment can be imposed. Some of the best fighting officers I was under were ignorant and degraded, unworthy associates in peace, but heroic, obedient, and faithful in performance of duty at the cannon's mouth.

When the 11th was organized, Col. Blaisdell was made Lieutenant Colonel at the suggestion of Col. George Clark, Jr., because he had been twelve years in the regular army and served in campaigns in Mexico, and was acquainted with the care and treatment of soldiers in war. Grant and Blaisdell were eight years in the same company, 4th U.S. Infantry, Blaisdell being a Sergeant and Grant a Lieutenant. When the fame of Grant rose above the military horizon in the West, Col. Blaisdell stated often these facts and also, that in the assault on a church in one of the last battles prior to the capture of the City of Mexico, he saved Grant's life. The door was broken open leaving a crevice of two inches and a Mexican was standing inside and raising his gun to shoot Grant when Blaisdell cut him down with his Sergeant's sword.

When Gen. Grant assumed the command of the Army of the Potomac, he sent for Col. Blaisdell and put him in command of the Irish Brigade as a temporary Brigadier General, and he was killed at Petersburg, June 23, 1864, defending his country and reflecting honor on the State of Massachusetts.

Until the discovery of Blake's unpublished memoirs, Blaisdell was known to historians only through the letters of Col.

Robert McAllister of the 11th New Jersey, which was brigaded with the 11th Massachusetts for much of the war. Blaisdell and McAllister were like vinegar and water: McAllister the highly religious temperance advocate, Blaisdell the coarse regular army veteran. The fact that Blaisdell was the senior officer of the two made McAllister dislike Blaisdell even more. McAllister's references to Blaisdell in his letters to his wife are uniformly negative. He was particularly troubled when Blaisdell asserted his right to command of the brigade in the winter of 1863-64, and was gleeful when several members of the 11th Massachusetts were charged with mutiny over the abolition of the Third Corps. In a letter dated March 31, 1864, McAllister wrote:

> *The exciting scene of yesterday was when an order came down ... to arrest and take under guard the officers under Col. Blaisdell who, at his insistence, held the indignation meeting. ... I am only sorry that the old chief of trouble and contention was not among the number taken. He is the man that ought to be caught. He has tried to do more to injure me this winter than any officer ever did. Yes, I wish Blaisdell was among them. They have only caught his tools.*[4]

James I. Robertson, Jr., editor of the McAllister letters, states that "the lax discipline present when Blaisdell was in command contributed highly to his failure to receive a brigadier's promotion." He cites OR LI/1, 1040-1041 to back up his charge. However, that citation leads to General Sickles' May 14, 1863 letter recommending Blaisdell for promotion: "[Blaisdell] has served with distinction and fidelity in all campaigns of this army, and is an officer of great merit." Moreover, Blake stated in his

---

4   For a detailed discussion of the courts-martial of the "mutineers," see *The Mutiny at Brandy Station.*

memoirs that Gen. Hancock told him Blaisdell was known as "Old Cruelty" because he was a strict disciplinarian. Robertson's book does further injury to Blaisdell's memory by inadvertently swapping the names under the photos of Blaisdell and General Alfred Torbert. Henry Blake's memoir sheds some light on why Blaisdell did not receive his general's star prior to his death:

> *Generals Hooker, Sickles, and Hancock recommended Blaisdell for promotion,[5] but Senator Sumner and Gov. Andrew were hostile because he was coarse and profane in language, ignorant and discourteous. When a leading politician urged Lincoln to remove Grant and submitted cogent reasons for his request, the President listened patiently and said, "I like this man Grant, he fights." This is a vindication of Blaisdell.*

Blaisdell was finally assigned to his own brigade command, still at the rank of colonel, during the siege of Petersburg, when he took over the "Corcoran Legion" (4/2/II), consisting of the 155th, 164th, 170th, and 182nd New York regiments. The circumstances of Blaisdell's demise were described as follows:

> *On June 23, Col. William Blaisdell was killed, being at the time on detached duty in command of the Corcoran Legion, Irish Brigade. While directing the construction of earthworks, he was dismounted, standing upon the raised mound of earth which subsequently became a portion of the main works of the Union army. While thus exposed, he was picked off by a rebel sharpshooter from the window of a house within the enemy's lines.*
> [Hutchinson]

---

5 Also General Prince, for Blaisdell's handling of the brigade at Mine Run.

*His negro orderly crawled out under fire and brought
his body back but life was extinct.* [Blaisdell Papers,
June 1948]

By the time the siege of Petersburg, Virginia, ended, ten months later, the army had constructed over thirty earth and timber "namesake" forts – to honor fallen officers – including a Fort Blaisdell, which was part of the secondary Union defense line near Jerusalem Plank Road.

A tall obelisk marks Blaisdell's grave at Hillside Cemetery, East Kingston, New Hampshire. The lengthy inscription lists the key dates of his life and military career, concluding with his posthumous promotion: "Brevetted Brigadier General for gallant and distinguished conduct, June 23rd, 1864, The first officer so honored during the Great Rebellion."

An article about Blaisdell in the family association's newsletter of June 1948 says that "An oil painting of him by a well-known Boston architect, Wm. Gibbons Preston, a gift from the painter, hangs on the wall at the head of the main stairway in the Cadet Armory." Preston was also the designer of that post-war building, which began its existence as the Armory of the First Corps of Cadets in 1897. The building survives to the present day, having been repurposed several times since the 1948 *Blaisdell Papers* article. The location of the oil painting, and William Blaisdell's connection to the Armory, remain unknown.

Photo credit: National Archives and Records Administration.
File 111-BA-428

Hooker at Williamsburg

May 5, 1862:

> *History will not be believed when it is told that the noble officers and men of my division were permitted to carry on this unequal struggle from morning until night, unaided, in the presence of more than 30,000 of their countrymen with arms in their hands! Nevertheless, it is true.* [Report of General Hooker, OR XI/1, 468]

Major Charles S. Wainwright, artillery chief for Hooker's division at Williamsburg, evaluated the performance of his commanders.

> *General Hooker was in the road just to the rear of my batteries, and under fire all day, as brave as the brave could be beyond doubt. But he seemed to know little of the ground where his infantry were fighting; and I must say did not impress me at all favorably as to his powers as a general. ... Soon after [came] the enemy's second attack, which took all our troops to drive back, say by 12:30 o'clock. General Hooker sent to Heintzelman and Sumner for reinforcements. None came until Kearny arrived, he says at two o'clock; I feel very sure it was nearly if not quite four o'clock. ... General Sumner had the whole of Gen. [Erasmus] Keyes' [Fourth] corps, three divisions, only a mile or two to our right. ... [Sumner] would not let Hancock go any further, and lay there all day with his 30,000 doing nothing. It is hard to understand. I know there is an awful amount of jealousy among our leading generals, but hardly think it can go so far; at the same time, it is equally hard to suppose they are actual fools.* [OR XI/1, 56-57]

General Sumner wrote a report defending against charges that he did not support Hooker:

I do not know what General Heintzelman means by asserting that three divisions were idle on the right during Hooker's engagement. There was but one division (Smith's) on Hooker's right, and from this Hancock was detached with his brigade, leaving but two brigades at the center until the arrival of General [John] Peck at the head of [Darius] Couch's division about 2 o'clock p.m., and shortly afterward the attack commenced on the center. [Silas] Casey's [3rd] division [Fourth corps] did not arrive on the ground till late in the afternoon. [Ibid, 452]

The last statement appears to be at odds with Casey's report that he was a mile and a half from the front at 10:30 A.M. and that he reported that fact to Sumner, who told him to get something to eat for his troops. Casey wrote:

About 1 o'clock p.m. I was ordered by General Keyes to advance to the front [up the Yorktown Road], and while making preparations so to do I was directed by General Sumner to move to the support of General Hooker, on the left. I immediately formed my division and moved off, with the first brigade [Naglee] leading, and gave directions for the other brigades to follow. After proceeding 3 miles I was overtaken by an express, directing me to obey the first order from General Keyes. I immediately countermarched and returned as quickly as possible. [Ibid, 557]

And Naglee (1/3/IV) reported that at 8 A.M. on the day of battle his brigade was not more than a mile and a half from the front and remained there five and a half hours until he received orders from Sumner to occupy "a position in the woods in front of his headquarters." Naglee continued:

Returning to execute this order, I informed General Casey of it, and he produced a written order from General Sumner directing the division to support General Hooker. This required it to move in the opposite direction. After a march of two hours, the artillery half the time up to the axles in the mud, the cavalry horses plunging, and the men in mud to their ankles, we arrived in the rear of General Hooker at 3:30 p.m., the very moment he was driven back by the enemy. We were preparing to support him, when another order came for all to countermarch and return in haste.

Meeting his corps commander, Keyes, on the way back, Naglee returned to Sumner's headquarters where McClellan had just arrived, and he (Naglee) was ordered at 4:30 p.m. to reinforce Hancock. As a true McClellan man, Naglee concluded: "The announcement of General McClellan's order was received in a manner than indicated but too plainly to his friends and our enemies the infinite pleasure his presence had immediately caused" [Ibid, 539]

Col. Regis de Trobriand, commanding the 55th New York in Peck's Brigade (3/1/IV) shed some light on who was where in the line of march on the morning of May 5:

Behind us marched General Kearny ... He urged on our stragglers, and told me that Hooker's division ... must have the whole rear guard of the enemy on his hands. Soon an aide to General Peck brought me the order to pass by Casey's division, which had halted, I do not know why, in a large open field, near a brick church. The sound of the cannonade did not diminish. At this point, Kearny's division turned to the left to come into line by a crossroad less encumbered.

A little further along, I met Captain Leavit-Hunt, aid to General Heintzelman, who had been ordered to hurry reinforcements. He informed me that the conjectures of Kearny as to Hooker were correct; ... In the absence of the general-in-chief, General Sumner and General Keyes lost time in consulting what was to be done. The former was senior in rank, but the latter alone had any troops within reach, and between the two, no measures were taken, and Hooker lost not only his position, but some of his guns. [Trobriand, 192-193]

The "infinite pleasure" of IV Corps leadership with the arrival of McClellan was not shared by the III Corps. McClellan had been absent at Yorktown to supervise the loading of Gen. William Franklin's (VI) Corps onto ships, to be landed farther up the peninsula. McClellan then exacerbated his negative image within the Third Corps when he reported to the newspapers how "Hancock, the superb" was responsible for the Union victory at Williamsburg. The III Corps did not consider either Hancock (or McClellan) the principal players at Williamsburg.

Martin L. Haynes wrote in his history of the 2nd New Hampshire (Grover's Brigade, Hooker's Division):

Hancock, in his little affair upon the right, – which many people in the regions where his regiments were raised still persist in believing to be the battle of Williamsburg, – lost but thirty-one men, a loss not equal to that of some companies in our division. [Haynes, 57]

On May 7, Major Wainwright echoed this view:

*...the division is feeling somewhat sore at all the glory
going to Hancock, who did very little fighting. It is said
that General McClellan has sent a second telegram to
Washington correcting his first, and complimenting
Hooker and his men. General Hooker is busy getting up
his report, and says that he means to see that his men
have all the credit they deserve.* [Wainwright, 60]

McClellan's "second telegram," of May 6, would appear to
have been equally disturbing to the III Corps divisions with the
only reference to Hooker being that his losses were "heavy."
Kearny's role was not mentioned at all. A couple of days later
Wainwright noted

*The ill feeling in [Hooker's] division on account of the
newspaper report of the Williamsburg battle continues
very strong. The more I think of it, the more credit our
men seem to me to deserve, but the less generalship I
see in Hooker, and more to blame in old Sumner and
perhaps Heintzelman too.* [Wainwright, 61]

McClellan's handling of Williamsburg in general, and
particularly the Hooker-Hancock episode, resulted in controversy
and hard feelings throughout the war. In late 1864 Regis de
Trobriand, a brigade commander in the remnants of the III Corps,
which was now the 3rd Division of Hancock's II Corps, pointed
out in a political discussion with Hancock when McClellan was
running for President, the latter's deficiencies in not giving some
credit to Hooker, Kearny, or Peck. Seeing that Hancock was
annoyed, de Trobriand added: "Not that I intend for a moment to
underrate in the least the importance of your part in the battle, in
that respect there can be but one voice, and, as much as anyone
can, I appreciate how much your brilliant action on that occasion

did you honor." Not impressed, Hancock ended the meeting stating "You are all alike in the old Third Corps. In your eyes, you have done everything in this war, and all others nothing." Ruefully de Trobriand concluded "I had not only lost the goodwill of my corps commander, but had also revived his prejudice against the whole Third division." [Trobriand, 659] Hancock, a Democrat serving the Republican Commander-in-Chief Abraham Lincoln, prudently did not publicly express his preference for President.

General McClellan portrayed himself to Secretary of War Edwin Stanton as the savior of the day at Williamsburg by appearing late in the afternoon at VI Corps headquarters:

Had I been one-half hour later on the field on the 5th [of May] we would have been routed and would have lost everything. Notwithstanding my positive orders I was informed on nothing that had occurred, and I went to the field of battle myself upon unofficial information that my presence was needed to avoid defeat. I found there confusion and incompetency, the utmost discouragement on the part of the men

He then demanded of Secretary Stanton to "return [the army] to the organization by division or else be authorized to relieve incompetent commanders of army corps." Lincoln and Stanton answered that "you may temporarily suspend that organization in the army now under your immediate command, and adopt any you see fit until further order." Lincoln wrote a surprisingly frank accompanying letter to McClellan reminding him that he (Lincoln) had ordered the reorganization on the unanimous recommendations of "twelve generals whom you have selected" and *on the unanimous opinion of every military man I could get an opinion from, and every modern military book, yourself excepted. Of course I did not on my judgment pretend to understand the subject. I now think it indispensable for you to know how your struggle against it is received in quarters which we cannot entirely disregard. It is looked upon as merely an effort to pamper one or two pets and persecute and degrade their supposed rivals. I have no word from Sumner, Heintzelman, or Keyes ... I am constantly told that you have no consultation or communication with them; that you consult and communicate with nobody but General Fitz John Porter and perhaps General Franklin. I do not say these complaints are true or just, but in all event it is proper you should know of their existence. Do the commanders of corps disobey your orders in anything?*

The President then posed the critical question:

Are you strong enough – even with my help – to set your foot upon the necks of Sumner, Heintzelman, and Keyes all at once? This is a practical and very serious question for you. [OR XI/3, 154-155]

The answer was "no." Although McClellan now had the authority and the backing from the President in writing, he did not exercise it. The corps system remained in place and their commanders continued in office.

Second Malvern Hill

By the end of July 1862 it was becoming apparent that the Army of the Potomac's month-long period of inactivity on the James River had to end with either a renewed advance toward Richmond or a complete withdrawal from the Peninsula. As a token military demonstration, Hooker's division was tasked with probing the Confederate forces before Richmond, ostensibly to make Lee think twice about McClellan's intentions, and keep Lee's focus on the Richmond front while Pope advanced in northern Virginia.

An article in *Frank Leslie's Illustrated Newspaper* of August 23, 1862 provides an example of field reporting that appeared in the popular press. It has no author or source attribution despite the use of the first-person singular in the text.

> *The Grand Reconnoissance to Malvern Hills*
> *August 5, 1862*
> *Malvern Hills have again been the scene of a*
> *conflict between the rebels and the National troops, and*
> *although not so glorious or sanguinary as was that of*
> *the 1st of July, it was no less conclusive on this point,*
> *that, when an army of National troops are led by a*
> *competent General, the Confederate tigers are no match*
> *for it. There is so much profound strategy about all the*
> *movements of the army on the James river, that it is*
> *difficult to decide what the real intention of the*
> *movement was; but the general opinion of the most*
> *intelligent officers on Hooker's staff is, that it was a*
> *feeler to ascertain the strength of the rebel army*
> *between Harrison's Landing and Richmond, in order to*
> *divert its attention from Pope and Sigel's advance. Gen.*
> *Hooker had arranged to make this reconnoissance on*

Saturday, the 2d of August, but the blunder or treachery of the farmer who had undertaken to guide the advance led to its postponement til Monday, the 4th of August, when at dusk all was in readiness to move.

The March

The column was not all upon the road until after the moon had risen, and threw a flood of light into every opening in the forests through which the little army passed. The roads were smooth and even, and the movement of the artillery train made but very little noise. Precautions were taken against alarming the enemy's pickets, one of whom – the only one on the road on which the column moved – was captured. Guards were placed around the houses on the way to prevent the inmates from conveying information of the approach of our troops. When the resting place for the night was reached, officers gave their orders in a whisper, as it was known that a camp of the enemy was near. The silence with which the affair was conducted would have been complete had it not been for a raw Brig.-Gen., who gave orders for brigade movements in a voice that could be heard a great distance in the stillness of the night.

The allusion to a "raw Brig-Gen." is obscure. It may refer to Col. Nelson Taylor, who wrote a report on the engagement as "commanding Second Brigade" [OR XI/2, 952] (Gen. Sickles was on leave.) Taylor had limited military service in the 1840's, but was a career lawyer in the prewar years. The other brigade commanders in Hooker's division were Cuvier Grover and Francis Patterson, both with extensive (and recent) military backgrounds.

The result of this little inflation of military pride shows how inexperience and lack of judgment in one officer, in a long line, may sometimes defeat or damage the success of an important movement. The result in this case was that the enemy were apprised of our approach at 12 o'clock, and reinforcements sent for. The bivouac for the night was in the rear of the Glendale battle-ground of June 30th, on the Quaker road... Grover's brigade was in advance, and most admirably were all their arrangements carried out.

Hooker's reconnaissance to Malvern Hill followed a round-about course from Harrison's Landing via Glendale, then making a left-about turn to approach Malvern Hill from the north.

At daylight of Tuesday, the 5th August, the column passed by Willis Church, the scene of the conflict where Heintzelman and Hooker formerly repulsed the rebels. The little church bore evidence of the bloody work of that day. The pews had all been torn out and converted into amputating tables and other conveniences for wounded men, and the floor of the building is entirely covered with blood. Many of the dead are buried around it, with hardly a memorial to designate them. I noticed on a headboard of one grave the name of L. E. L. Blane, 3d Louisiana regiment.

Not to be confused with the 3rd Louisiana regiment that fought entirely in the western theater, the "3d Louisiana regiment" present at Glendale (Frayser's Farm) was a battalion of eight companies which had originally been recruited as the "2nd Regiment, Polish Brigade" at Camp Pulaski, Amite City, Louisiana. By early August 1862 it had been combined with two companies of the 7th Louisiana battalion to form the 15th

Louisiana regiment, which went on to fight in every eastern theater campaign for the duration of the war. Online rosters of the 15th Louisiana do not include members who died before the amalgamation of its constituent battalions.[6]

> *In the woods, beyond the church, on the road to Malvern Hills, are the abundant evidences of the artillery practice of the enemy upon the rear of our army during the march to James river. Trees of considerable size are literally mowed down by the shot and shell, and others scarred by Minié bullets that hailed through them.*
>
> *The column arrived at Malvern Hills about half-past five o'clock. The sky was clouded and lowering and a heavy moisture hung in the air, which was not stirred by the slightest breeze. It is not proper to state what force comprised the expedition. Suffice to say that it was a strong and well proportioned corps, with the proper amount of cavalry and artillery.*
>
> *Cavalry and artillery led the column into the field, through the wooded gorge in front of Malvern Hills, accompanied by Gen. Hooker and his staff. Taking a position, with his staff, upon a conspicuous knoll, Gen. Hooker opened the fight in person, as he did at Williamsburg, posting his artillery and ordering the cavalry to the attack. The enemy had a battery in position at the right of the brick house on the edge of the bluff towards James river. This was the Fauquier battery, from Fauquier county, Virginia, commanded by Capt. [Robert M.] Stribling. The 8th and 17th Georgia regiments were on the field and two regiments of*

---

6 From acadiansingray.com January 2026.

*cavalry – one the Cobb cavalry, of Georgia. Major
Pickett was in command of the post.*

Possibly Maj. John H. Pickett, a 17th Georgia field officer.
[NPS Soldier Database] The identities of specific rebel regiments
were apparently obtained from the Confederate prisoners.

*The 8th Illinois cavalry[7] charge upon the battery,
and were received by a storm of grape and canister that
drove them back. The battery then fired shell and
spherical case directly, and Capt. Benson's battery
replied.[8] Almost instantly a cloud of smoke settled over
the field, obscuring every object. Then enemy kept up a
remarkably well-directed fire upon the road, on which
or infantry were advancing. Shell and spherical case
burst in the road and upon either side of it in rapid
succession. The enemy's practice showed that they had
previously obtained the range of this spot with great
accuracy. The infantry, led by the 1st brigade, Hooker's
division, commanded by Gen. Grover, marched up the
road under this fire without flinching, like veterans, in
admirable order, followed by the other brigades in the
same manner. When a shell was heard approaching a
regiment the men dropped, and were instantly on their
feet again, with unbroken ranks, moving forward.
Fortunately only two or three shells took effect upon
the column, killing two men and wounding 15. The
infantry marched upon the field and took their position,
but as the smoke had become so dense that is was
impossible to see the enemy, Gen. Hooker ordered the*

---

7   Led by Col. John F. Farnsworth, who was promoted to brigade command
in time for the Antietam campaign.

8   Capt. Henry Benson led Battery M, 2nd U.S. horse artillery. He died
August 11, 1862 from wounds received at 2d Malvern Hill.

fire of the artillery to be slackened. When the smoke cleared away, it was discovered that the enemy was retiring, and when Hooker resumed the fire, their retreat was considerably quickened.

It was about this point in the engagement that the reserve brigade (3/2/III), presumably under command of Brig. Gen. Francis Patterson, was ordered by Hooker to cut off the retreating rebels, and failed to do so – the reason for which Blake caustically states in Chapter VI. Patterson had been ill with typhoid fever during June and July 1862, and there is some uncertainly about exactly when he resumed his duties. But given that 2nd New Hampshire historian Martin Haynes identifies Patterson by name in his 1865 book, and Henry Blake's allusions are unambiguous, it seems that Patterson was back in command of his brigade by early August.

For two miles did our gallant men pursue the disorganized rebels, who only were saved by the closing in of the night from a heavy loss in prisoners
Our men remained masters of the field of battle, where they bivouacked for the night, and where they remained until Friday, when the object of their reconnoissance having been fully accomplished, they fell back to their previous position. It is only proper to state that, when the battle was over, Gen. McClellan arrived on the field, and congratulated Gen. Hooker, with much cordiality, upon his success. Considering the nature of the engagement, the casualties on both sides were very small, the rebels greatest loss being in prisoners, of whom about 100 were captured.

Grover at Groveton

On the morning of August 29, 1862, the brigade of Gen. Cuvier Grover forded Bull Run and marched over the battlefield of the previous year. That morning, Schurz' Division of Sigel's Corps had briefly occupied a portion of the unfinished railroad, and had been driven off by A. P. Hill's Confederate division. The Union division of Kearny (1/III) was supposed to support Schurz by attacking A. P. Hill on Schurz' right, but Kearny was not cooperating because of a personal animosity toward Sigel. The latter was eventually reinforced by two regiments of the Jersey Brigade (3/2/III), and two regiments from the IX Corps. Further to the left was Gen. Robert Milroy's independent brigade of Sigel's Corps, which had attacked Lawton's Brigade (Ewell's Division) at the unfinished railroad, but had been badly battered and fallen back to Groveton woods. In the early afternoon, Grover's Brigade was dispatched to reinforce Milroy and was temporarily placed under command of Sigel.

Pope's battle plan on August 29 was for the corps of Porter and McDowell to attack the right of Jackson's line along the unfinished railroad. But Pope was not aware that Longstreet had moved to the same position about 1:00 P.M. While waiting for Porter's attack to begin, Pope decided to initiate some attacks on Jackson's front, presumably to distract Jackson's attention from the threat to his right. These attacks were to made by Grover's Brigade on the left and Robinson's Brigade (Kearny's Division) on the right.

Before making their attack, Grover's Brigade had marched past Starke's and Lawton's position at the unfinished railroad and formed for attack opposite the gap between A. P. Hill's brigades of Gregg and Thomas. The exact nature of this movement is somewhat speculative: neither Blaisdell nor Grover mentioned a preliminary shift of front in their reports. Grover declared that his reconnaissance of the Confederate position was at the time of his arrival on the field, and that no change of direction was ordered after the brigade's initial forward movement through the woods in front of the Confederate line. Milroy maintained that he told Grover where to direct his attack:

> *I saw him forming a line to attack the rebel stronghold in the same place I had been all day, and advised him to form a line more to the left, and charge [with] bayonets upon arriving at the railroad track.* [OR XII/2, 320]

Grover explained the simple strategy for the charge, and the result.

> *Instructions were given for the line to move upon the enemy until it felt his fire, then close upon him rapidly, fire one well-directed volley, and rely upon the bayonet to secure the position on the other side.*
>
> *We rapidly and firmly pressed upon the embankment, and here occurred a short, sharp, and obstinate hand-to-hand conflict with the bayonet and clubbed muskets. Many of the enemy were bayoneted in their tracks, others struck down with the butts of pieces, and onward pressed our line. In a few yards more it met a terrible fire from a second line which in its turn broke. The enemy's third line now bore down upon our thinned ranks in close order, and swept back the right center and a portion of our left. With the gallant Sixteenth Massachusetts on our left I tried to turn his flank, but the breaking of our right and center and the weight of the enemy's line caused the necessity of falling back, first to the embankment and then to our first position, behind which we rallied to our colors.* [Ibid, 439]

Col. Blaisdell described the charge:

Driving the enemy's picket before it, the regiment came upon and engaged a heavy line of the enemy's infantry, which was driven back and over a line of railroad where the roadbed was 10 feet high, behind which was posted another heavy line of infantry, which opened a terrific fire upon the regiment as it emerged from the woods. The Eleventh Regiment, being the battalion of direction, was the first to reach the railroad, and of course received the heaviest of the enemy's fire. This staggered the men a little, but, recovering in an instant, they gave a wild hurrah and over they went, mounting the embankment, driving everything before them at the point of the bayonet. Here for two or three minutes the struggle was very severe, the combatants exchanging shots almost muzzle to muzzle and engaging hand-to-hand in deadly encounter. Private John Lawler, of Company D, stove in the skull of one rebel with the butt of his musket and killed another with his bayonet. The enemy broke in confusion and ran, numbers throwing away their muskets, some fully cocked and the owners to much frightened to fire them, the regiment pursuing them some 80 yards into the woods, where it was met by an overwhelming force in front, at the same time receiving artillery fire which enfiladed our left and forced it to retire, leaving the dead and many of the wounded where they fell. It was near the railroad embankment that the brave Tileston, Stone, and Porter, and other gallant men received their mortal wounds.

Being thus overpowered by numerical odds, after breaking through and scattering two lines of the enemy and compelled to evacuate the woods and enter into the open fields beyond, the enemy pursuing us hotly to the edge of the woods, I was greatly amazed to find that the regiment had been sent to engage a force of more than five times its numbers, strongly posted in thick woods and behind heavy embankments, and not a soldier to support in case of disaster. [Ibid, 441]

The numerical odds were more on the order of three-to-one. Grover had struck the Confederate line in the gap between two brigades: Gregg and Thomas. It was Pender's Brigade that was the "overwhelming force" thrown in at a crucial stage of the fighting that forced Grover's Brigade to fall back.

As to the three officer deaths mentioned by Blaisdell: Lt. Col. George F. Tileston had been married only eight months, making his pregnant wife back home the first war widow in Duxbury; and Capt. Benjamin Stone of Blake's Company K lay on the field four days before being taken to a hospital in Washington where he died September 10th following a leg amputation.

Lt. William Rogers Porter, 21, of Dorchester, had been promoted to Lieutenant from the ranks of the 13th Massachusetts, joining the 11th regiment in January 1862. His death was conveyed to his family by letter from Col. Blaisdell:

Your brother...died on the field of battle at Bull Run...while gallantly leading his company in a bayonet charge upon a very strong position of the enemy, who were posted behind a railroad embankment.

Before reaching the railroad, Lieut. Teaffe observed him suddenly bend forward, and totter; and caught him in his arms, and asked him if he was hurt. He replied, "My back is broken." In answer to a second question, as to where he was hit, he said he was shot in the bowels.

*By order of Lieut. Teaffe, Sergeant Farrington,
assisted by Sergeant Boucher, bore him a short distance
to the rear, where he died. Sergeant Boucher, in a few
minutes after, having returned to the battle, was killed.*

*I cannot omit this opportunity to do justice to the
memory of your brother, who, during the short time he
was an officer in the regiment, performed his duty
understandingly, faithfully, and fearlessly; winning the
love of his men – who felt the greatest confidence in him
as a leader – the esteem of his brother officers, and the
respect of all who were acquainted with him. I feel I
have lost a valuable and promising officer, and
sympathize deeply with you and his large circle of
friends at home in their bereavement, trusting you and
they may find consolation in the knowledge that the
manner of his death, in the van of battle for human
freedom, and the perpetuity of our institutions, formed a
fitting termination to the life of a worthy and brave
soldier.* [Porter Family]

Porter's family recovered his body later that same year:

*They went at just the right time; they found,
unexpectedly, just the right man to lead them to the spot
where it had been needful for grieving comrades to
abandon their dead. There, undisturbed, had he been
lying; the withering leaves had fallen, and the autumn
winds were sighing above his grave, but no profane
hand, no careless foot of man or beast had violated its
peace. The marks of identity were such as to leave no
possibility of any painful doubt.* [Ibid.]

For Brigadier General Cuvier Grover, the bayonet charge at
2nd Bull Run would be the brilliant moment of his army career.

Gen. Heintzelman (III Corps commander) deemed it "the most gallant and determined bayonet charge of the war." [OR XII/2, 413] *Harper's Weekly* published a profile of Grover, November 12, 1864:

> *General Grover is a native of Maine, and graduated at West Point in 1850. At the beginning of the war he had for some time been performing military duty in Utah, but immediately he was recalled and assigned to a command in the Army of the Potomac. He commanded a brigade under General Hooker in the Peninsular Campaign. In the latter part of [1862] he was transferred to the Department of the Gulf, where he commanded a division of the Nineteenth Corps, [which] participated in the siege of Port Hudson. In July [1864] the Nineteenth Corps was recalled to Virginia, and General Grover…was wounded on [October] 19th, while leading a charge…in the battle of Cedar Creek.*

After the war Grover continued his military career with the U.S. Cavalry on the western frontier. He died in 1885 in Atlantic City, New Jersey.

Mott at Spotsylvania

On the afternoon of the May 10, 1864 a supposedly coordinated offensive was planned for about 5:00 P.M. against the "Mule Shoe" salient north of Spotsylvania Court House. The VI Corps was to attack on the west side of the salient while Mott's Division (4/II) supported at the apex of the salient about a quarter mile to the north. However, the starting time for the "coordinated" attack seems to have been in flux, confused by the V Corps assault at Laurel Hill earlier in the afternoon. VI Corps commander Horatio Wright, in overall command of the assault, originally instructed Mott to attack at 5:00 P.M. and later changed this to 6:00, but there is no evidence that this change was ever communicated to Mott. Both Blake and Lt. John Schoonover (11th New Jersey) declared that the time of the charge was 5:30, which is not that inconsistent with McAllister, who stated that they started through the woods driving skirmishers at 5:00, bringing them to the open field at a time closer to 5:30.

Earlier Mott had been ordered to link up with Burnside's IX Corps to the east and support them if they were attacked. To do this Mott had dispatched an undetermined portion of the Excelsior Brigade to find them: "Apparently most of Mott's Second Brigade, the old Excelsior, had been assigned to locate Burnside's left." [Matter, 161] However, the 11th Massachusetts and the 73rd New York of the Excelsior brigade were definitely a part of the May 10th charge.

Mott notified Wright at 2:05 P.M. that because of his extended picket line toward Burnside he would have only 1,200 to 1,500 men in the assault, including two supporting regiments from the VI Corps. Mott asked whether he should "call in the pickets on my left" to augment his forces. Wright answered that if Mott could not get his pickets back at 5:00 go ahead with the attack. [OR XXXVI/2, 603]

Col. Robert McAllister, 1/4/II commander, was in charge of the assault while Gen. Mott apparently remained in his headquarters at Brown house, about three-fourths of a mile away from the point of attack. McAllister reported that with the 6th New Jersey as skirmishers:

> *General Mott instructed me forward at 5:00 P.M., my brigade in front line, Second [Excelsior] Brigade in rear, Colonel [Edward L.] Campbell on my right with two regiments [1st and 15th New Jersey]. We moved through the woods and drove the enemy's skirmishers back towards their works. On reaching the open field, the enemy opened his batteries, enfilading our line and causing our men to fall back in confusion, excepting a small portion of the front line. Colonel Blaisdell, Colonel Campbell, and myself consulted as to what was to be done, and concluded there was nothing left but to fall back, which we did, to the foot of the hill. Before reaching this place we threw out a line of pickets in advance of the old one, and massed our forces as a reserve. [OR XXXVI/1, 490]*

The division had encountered Edward Johnson's Division of Ewell's Corps (their opponents at Locust Grove the previous November during the Mine Run Campaign), along with twenty-two pieces of artillery.

In contrast, the initial phase of the VI Corps attack was a great success, resulting in the promotion its leader, Col. Emory Upton, to brigadier general. Compared to Mott's attack, Upton attacked with probably twice as many men (twelve regiments), there was genuine surprise, and most importantly, the open ground which had to be covered was much less. Upton was within 200 to 250 yards while Mott's troops had to charge five times that distance. Matter notes that "It probably took the three regiments of the first line [of Upton] no longer than 60 to 90 seconds to reach the [Confederate] line." [Matter, 160-162] In fact Upton's attack had a great deal in common with that of Grover's Brigade at 2nd Bull Run in the element of surprise and the relatively short distance required for a bayonet charge. Union army chief of staff Humphreys wrote:

> *Upton's attack was concealed from their view and was a*
> *surprise, and the plan of assault being well arranged*
> *and carried out, was a success. The plan and manner of*
> *Mott's assault, on the contrary, did not admit of being a*
> *surprise. The formation of his troops probably kept the*
> *attention of the enemy upon him, and in that way helped*
> *more effectually to conceal Upton's preparations.*
> [Humphreys VC, 86-87]

Two days later (May 12) Grant, encouraged by Upton's success, ordered a dawn attack by the entire II Corps against the apex of the "Mule Shoe." The assault succeeded in capturing the entrenchments and the bulk of Johnson's Division (including Generals Johnson and Steuart). The success of the attack has been attributed to the fact that the twenty-two artillery pieces Mott had faced on May 10th had been moved to another sector of the line. However, stealth and numerical superiority played a large part as well.

Henry Blake was wounded on the morning of May 12 in the interval between the capture of the "Mule Shoe" and the Confederate counterattack that resulted in the fight for the "Bloody Angle." The two armies squared off on opposite sides of the same breastwork for the remainder of the day. Captain

Charles C. Rivers (Company A) stated that after the rebel counterattack:

> *...the line of works was soon reversed; the fighting now became desperate on both sides, and continued without cessation until dark, with a loss to the regiment of five killed, two commissioned officers and thirty-three enlisted men wounded, and four missing supposed to have been taken prisoner.* [Hutchinson, 64]

The II Corps line at the apex of the salient, after the Confederate counterattack, was held by the divisions of Barlow, Birney, and Mott, left to right. In his history of II Corps Walker wrote:

> *The division commanders of the corps, and the brigade commanders, won new honors at the bloody salient. Among regimental commanders Colonel William Blaisdell, of the Eleventh Massachusetts, deserves especial mention for unflinching determination in holding his line against the most desperate assaults.* [Walker, 479]

Actually, Blaisdell was in command of the brigade through the Spotsylvania campaign in the absence of Col. William Brewster, who had taken sick leave during the Wilderness – a point also overlooked by Humphreys in his 1883 book *The Virginia Campaign* (pp. 86, 94).

Matter wrote that although Mott's Division was regarded "as the weak link in the chain of the Federal Second Corps, ... many of them fought very well indeed that day [May 12]." He pointed particularly to Col. Napoleon B. McLaughlen volunteering the 1st Massachusetts, whose enlistment expired within a week, to replace Upton's Brigade on the west face of the salient, and Lt-

Col. Waldo Merriam who was killed leading the 16th Massachusetts. [Matter, 253-254]

Because of its lackluster performance on the first day of the Wilderness, and its abortive charge on May 10, Mott's Division cemented its reputation as the "weak sister" of the II Corps. On May 11 Mott himself gave a pep talk to his troops in the form of a special order:

> *The brigadier-general commanding is grieved to notice that the Fourth Division, Second Corps, is sacrificing the reputation of 'Hooker's old Division' ... There is no excuse for such conduct. ... The time calls for greater exertion than any previous one, and the commanding general expects it will be put forth. Commanding officers of brigades, regiments, and companies are called upon and enjoined to exercise that example and authority required of them by existing orders and regulations. Let us show to the army and the world that our part has been fully and faithfully performed.* [OR XXXVI/2, 637]

Mott's Division fought creditably on the 12th but on the 13th it was eliminated as a division and constituted as the 3rd and 4th Brigades of Birney's 3rd Division, II Corps. [Ibid. 711-712] The expressed rationale was its depleted strength from battle losses, and the impending expiration of the terms of enlistment of a number of it regiments. As it turned out, this reasoning spared the "old III Corps" from the Cold Harbor assaults of early June.

Negative comments on Mott and the 4th Division come from headquarters staff and persons they came in contact with. Lt.-Col. Theodore Lyman on Meade's staff pointed to "discontent about breaking up of the old III Corps" and to recently being under "commanders of little merit or force of character." After Spotsylvania Lyman wrote:

> *Mott's division, that had hitherto behaved so badly, was broken up and put with Birney...a sad record for Hooker's fighting men!* [Lyman, 93, 114]

The "old III Corps" had acquired a reputation as grumblers in the Army of the Potomac dating back to their very first battle at Williamsburg, May 5, 1862. Hooker had complained that his division was heavily engaged all day – "unaided in the presence of more than 30,000 of their countrymen" – while Hancock's flanking move was magnified out of proportion by McClellan and the press, who championed "Hancock the superb." The III Corps' reputation only got worse under Sickles (at Chancellorsville and Gettysburg) and French (at Mine Run) despite significant participation in all Eastern theater campaigns except Antietam. Finally, in March 1864 the III Corps was *"killed by General Order No. 9 – Actuated by Personal Malice, Spite and Jealousy."* So said the epitaph on a wooden gravestone for the corps, set up in the camp of the 63rd Pennsylvania. [Hays, 224] The members of the III Corps felt that their sacrifices of the previous two years had been marginalized by the breakup of the corps.

Bibliography

Adams, George W. *Doctors in Blue: The Medical History of the Union Army in the Civil War*, Baton Rouge : 1996.

Adams, William Taylor (Oliver Optic). *The Young Lieutenant*, Boston: 1865. Juvenile novel about the battle of Williamsburg.

Armstrong, William Henry. *Red-Tape and Pigeon-Hole Generals*. Charlottesville, VA : 1999. Commentary by Frederick B. Arner. **(Red Tape)**

Arner, Frederick B. *The Mutiny at Brandy Station: The Last Battle of the Hooker Brigade*, Kensington, MD : 1993. **(Mutiny)**

Baldwin, Clark B. Letter to Batchelder, May 20, 1865; First Massachusetts Infantry Veterans Association, "Where Honor Is Due"

Blaisdell, Nathaniel. "General William E. Blaisdell," *Blaisdell Papers*, June 1948, Vol 3, No 5. The Blaisdell Family Association.

Blake, Henry Nichols. "Familiar Quotations in War," unpublished address to the Milton Historical Society, October 4, 1916. **(Familiar Quotations)**

————. Memoirs, unpublished manuscript, 1916. **(Memoirs)**

————. "The Personal Record of a Civil War Veteran, Montana Territorial Editor, Attorney, Jurist," *Montana: The Magazine of Western History*, October 1964, Volume 14, Number 4. Excerpts from Blake's memoirs, edited by Vivian Paladin. **(Montana)**

————. "Tales by an Old Harvard Soldier," Harvard Alumni Bulletin, June 16, 1927. **(Old Soldier)**

————. *Three Years in the Army of the Potomac*, Boston : 1865.

Bowen, James L. *Massachusetts in the War,* Springfield : 1889.

Busey, John and Martin, David G. *Regimental Strengths and Losses at Gettysburg*, Hightstown, NJ : 1986. **(RSL)**

Cable, Mary. *The Avenue of Presidents*, Boston : 1969.

Cooper, Thomas V. "26th Pennsylvania Volunteer Infantry, Pennsylvania's Memorial Days," Library of Congress, 1889. Manuscript.

Coughenour, Kevin. "Andrew Atkinson Humphreys: Divisional Command in the Army of the Potomac," Sixth Annual Gettysburg Seminar, 1998.

Cudworth, Warren H. *History of the First Regiment (Massachusetts)*, Boston : 1866.

Cullum, George Washington. *Biographical Register of the Officers and Graduates of the United States Military Academy*, U.S. Military Academy : 1866 and later editions.

Davis, William C. *Battle at Bull Run*, Mechanicsburg, PA : 1995.

Denney, Robert E. *Civil War Prisons and Escapes*, New York : 1993.

Dyer, Frederick. *Compendium of the War of the Rebellion*, New York : 1959.

Fishel, Edwin C. *The Secret War for the Union*, Boston : 1996.

Ford, Henry E. *History of the 101st Regiment*, Syracuse, NY : 1898. 101st New York.

Fox, William F. *Regimental Losses In The American Civil War*, Albany, N.Y. : 1889.

Gallagher, Gary, ed. *The Fredericksburg Campaign*, Chapel Hill, N.C. : 1995.

Gazet, Edward K. Letters and pencil sketches. Collection of James F. Catania, Jr.

Graham, Martin F. and Skoch, George F. *Mine Run: A Campaign of Lost Opportunities*, Lynchburg, VA : 1987. **(Mine Run)**

Grant, Ulysses, S. *Personal Memoirs of U. S. Grant*, New York, 1885.

Hains, Peter C. "The First Gun at Bull Run," *The Cosmopolitan* (magazine), Volume 51, 1911.

Hall, Clark B. "Season of Change: The Winter Encampment of the Army of the Potomac," *Blue and Gray Magazine*, April 1991.

Hamilton, Edith. *Mythology, Timeless Tales of Gods and Heroes*, New York, NY : 1999.

Hastings, Earl C. and David. *A Pitiless Rain, The Battle of Williamsburg*, Shippensburg, PA : 1997.

Haswell, Charles H. *Reminiscences of New York by an Octogenarian (1816-1860),* New York : 1896.

Hays, Gilbert. *Under the Red Patch,* Pittsburgh : 1908. 63rd Pennsylvania.

Haynes, Martin. *History of the Second Regiment New Hampshire Volunteers: Its Camps, Marches, and Battles.* Manchester, NH : 1865.

————. *A History of the Second Regiment New Hampshire Volunteers in the War of the Rebellion.* Lakeport, NH : 1896. (Reprint and expansion of the 1865 edition.)

Henderson, William D. *The Road to Bristoe Station*, Lynchburg, VA : 1987.

Hennessy, John J. *The First Battle of Manassas*, Lynchburg, VA : 1989.

————. *Return to Bull Run*, New York : 1993. **(Hennessy 1993)**

Higginson, Thomas W. *Massachusetts in the Army and Navy during the War of 1861-65,* Boston : 1895/6.

Historical Datasystems, Inc. Civil War Database, Kingston, MA. Online database. **(HDS)**

Holland, Mary Gardner. *Our Army Nurses*, Roseville, Minn.: 1998.

Holmes, Oliver Wendell. *Touched With Fire, Civil War Letters and Diary*, Cambridge : 1946.

Hudson, Carson O. *Civil War Williamsburg*, Mechanicsburg, PA : 1997.

Humphreys, Andrew A. *From Gettysburg to the Rapidan*, New York : 1883. **(Humphreys GR)**

————. *The Virginia Campaign, 1864-5*, New York : 1995. Facsimile reprint of 1883 edition. **(Humphreys VC)**

Humphreys, Henry. *Andrew Atkinson Humphreys: A Biography*, New York : 1924. **(Humphreys Bio)**

Hutchinson, Gustavus B. *A Narrative of the Formation and Services of the Eleventh Massachusetts*, Boston : 1893.

Johnson, Robert Underwood, ed. *Battles and Leaders of the Civil War*, New York : 1888. **(B&L)**

Kirsch, George B. *Baseball in Blue and Gray*, Princeton, NJ : 2007.

Lee, Richard M. *Mr. Lincoln's City*, McLean, VA : 1981.

Leech, Margaret. *Reveille in Washington*, New York : 1941.

Leslie, Frank. *Frank Leslie's Illustrated Newspaper*, New York : 1862.

Lonn, Ella. *Desertion During The Civil War*, Lincoln, NE : 1998. Reprint of 1928 edition.

Lord, Francis A. *They Fought for the Union*, Harrisburg, PA : 1960.

Lowry, Thomas P. *The Story the Soldiers Wouldn't Tell*, Mechanicsburg, PA : 1994.

Lyman, Theodore. *Meade's Headquarters 1863-1865, Letters of Colonel Theodore Lyman*, Boston : 1922. Edited by George R. Agassiz.

Marbaker, Thomas D. *History of the Eleventh New Jersey Volunteers*, Trenton : 1898. Reprinted 1990, edited by John W. Kuhl.

Massachusetts Adjutant General. *Annual Report for 1862*, Boston : 1863.

————. *Annual Report for 1864*, Boston : 1865.

————. *Massachusetts Soldiers, Sailors, and Marines in the Civil War*, Norwood, MA : 1931. Eight volumes. **(MSSM)**

Matter, William D. *If It Takes All Summer: The Battle of Spotsylvania*, Chapel Hill, NC : 1988.

McAllister, Robert. *The Civil War Letters of General Robert McAllister*, Baton Rouge : 1998. Edited by James I. Robertson, Jr.

Military Order of the Loyal Legion of the United States, Massachusetts Commandery. **(MOLLUS)**

Morss, Walter S. Diaries and letters. Collection of Stanwood R. Morss.

National Archives and Records Administration. **(NARA)**

National Park Service. *Soldiers and Sailors Database.* nps.gov/civilwar/soldiers-and-sailors-database.htm **(NPS)**

Pearcy, Matthew T. "Andrew A. Humphreys at Chancellorsville and Gettysburg" *Army History* (magazine), Summer 2013.

Perry, Milton F. *Infernal Machines*, Baton Rouge : 1985.

Pitch, Anthony S. *The Burning of Washington, The British Invasion of 1814*, Annapolis, MD : 1998.

Pfanz, Harry W. *Gettysburg: The Second Day*, Chapel Hill, NC : 1987.

Porter Family. *Memorial of William Rogers Porter*, privately published, 1862, for Porter's funeral. **(Porter Family)**

Posey, Calvert K. and Judith I. "A History of the Role of Charles County in the Civil War," Monograph.

Priest, John Michael. *Before Antietam: The Battle for South Mountain*, New York : 1996.

————. *Nowhere To Run: The Wilderness*, Shippensburg, PA : 1994.

Rhea, Gordon C. *The Battle of the Wilderness*, Baton Rouge: 1994.

————. *The Battles for Spotsylvania Court House and the Road to Yellow Tavern May 7-12, 1864*, Baton Rouge : 1997.

Sassaman, Richard. "The Washington Tragedy: Dan Sickles and the Trial of the (19th) Century." American History, Vol. 33, No. 4, October 1998.

Scott, Robert G. *Into the Wilderness with the Army of the Potomac*, Bloomington, IN : 1988.

Sears, Stephen W. *Chancellorsville*, New York : 1996. **(Sears 1996)**

————. *To The Gates of Richmond*, New York : 1988. **(Sears 1988)**

————. *The Young Napoleon*, New York : 1999. **(Sears 1999)**

Shreve, William P. *The Story of the Third Army Corps Union*, Boston : 1910. Mutual aid society of Third Corps officers.

Sifakis, Stewart. *Who Was Who in the Union*, New York : 1988.

Starr, Stephen Z. *The Union Cavalry in the Civil War*, Baton Rouge : 1979.

Steere, Edward. *The Wilderness Campaign*, Harrisburg, PA : 1960.

Stevens, Charles A. *Berdan's United States Sharpshooters in the Army of the Potomac*, Dayton, OH : 1972 (First published in 1893).

Trobriand, Régis de. *Four Years with the Army of the Potomac*, Boston : 1889.

Tyler, Mason Whiting, *Recollections of the Civil War*, New York : 1912.

The Union Army: A History of Military Affairs in the Loyal States, 1861-65, Madison, WI : 1908. Eight vols. **(Union Army)**

United States Army Military History Institute, Carlisle, PA. Unit bibliographies and an extensive collection of soldier photographs. **(USAMHI)**

United States War Department. *The War of the Rebellion: A Compilation of the Official Records of the Union and Confederate Armies*, Washington: 1881-1902. All citations are from Series I. **(OR)**

van Santvoord, Cornelius. *The One Hundred and Twentieth New York State Volunteers*, Saugerties, NY : 1997 (First published ca. 1894).

Wainwright, Charles S. *A Diary of Battle: The Personal Journals of Colonel Charles S. Wainwright*, New York : 1962. Edited by Allen Nevens.

Walker, Francis A. *A History of the Second Corps*, New York : 1887.

Wheat, Thomas Adrian. *A Guide to Civil War Yorktown*, Knoxville, TN : 1997.

Wheeler, William H. Papers. Duke University. Wheeler was an officer in both the 16th and 11th Massachusetts.

Wiley, Bell I. *The Life of Billy Yank*, Indianapolis : 1952.

Wills, Mary A. *The Confederate Blockade of Washington, D.C. 1861-1862*, Shippensburg, PA : 1998.

www.ingramcontent.com/pod-product-compliance
Lightning Source LLC
Chambersburg PA
CBHW071445140726
47997CB00005B/1600